LANDSCAPE ECOLOGY - THE INFLUENCES OF LAND USE AND ANTHROPOGENIC IMPACTS OF LANDSCAPE CREATION

Edited by **Amjad Almusaed**

Landscape Ecology - The Influences of Land Use and Anthropogenic Impacts of Landscape Creation

http://dx.doi.org/10.5772/61905
Edited by Amjad Almusaed

Contributors

Hadi Memarian, Siva Kumar Balasundram, Ignacio Melendez-Pastor, Encarni I. Hernández, Jose Navarro-Pedreño, Ignacio Gómez, Magaly Koch, Richard Malcolm Thackway, María-Luz Rodríguez-Blanco, Radoslava Kanianska, Gica Pehoiu, Mihaela Sencovici

CBS, Edition **2017**

Published by InTech

Janeza Trdine 9, 51000 Rijeka, Croatia

Publishing Process Manager Dajana Pemac

Technical Editor SPi Global

Cover Designer InTech Design Team

Additional hard copies can be obtained from orders@intechopen.com

Landscape Ecology - The Influences of Land Use and Anthropogenic Impacts of Landscape Creation
Edited by Amjad Almusaed
p. cm.
Print ISBN 978-953-51-2513-6
Online ISBN 978-953-51-2514-3

INTECH

open science | open minds

Contents

Preface

The term Landscape ecology is an area of ecology applied to dealing with the scientific study of the structure, function, dynamics, and changing ecosystems in a geographic scope (landscape) containing structural landscape heterogeneity, composed of many backgrounds of interaction based on the existing topography and soils. The landscape ecology is an applied science, created originally as an interface between geography and ecology. Landscape ecology studies the agricultural landscapes and nature, and the major impact of human activities such as agriculture, forestry or urban development on the landscape ecology, and economic development measures envisaged for the area while maintaining the existing natural ecosystems and their ecological communities unaltered. Landscape ecology has rapidly emerged in the past decade to become usable and relevant to practicing land-use planners and landscape architects. To focus on complex land mosaics, such as neighborhoods, whole landscapes, and districts, is at progressively the critical spatial scale. A significant contribution to the birth of landscape ecology is derived from studies of the vegetation and the mapping of the vegetation units. The cartography of the vegetation sets the stage for a performance of the environmental diversity of terrestrial ecosystems. In this representation of the environmental diversity, plants play a particularly prominent role. Nowadays, landscape ecology becomes valuable tools to stabilize the balancing of the land-use process of urban planning and design and maximize the role of landscape architects. If a society requirement is to create an urban element, such as street, water mirror, and special building functions, that is an actual act.

This book is for landscape specialists and those who have an interest in this area, where the ecology and the land-use and anthropogenic impacts of landscape creation, which will be illustrated by methods and applications, will be vital fragments of this research book. Also, this book addresses several very different subjects of study: ecological interpretation of land-use act, landscape district in applied ecological analysis, and anthropogenic impacts on landscape creation.

A book, which is a creation of act similar to a building idea, is always a concerted effort that consists of the input of many people over a long period. Therefore, I would like to express my sincere sense of appreciation and thankfulness to all the authors for their valuable contributions. At the same time, I would like to thank Ms. Dajana Pemac, InTech's Publishing Process Manager, for her assistance and efficiency in the management of this book and her cooperation at various phases of the book publication.

Amjad Almusaed
University of Basrah
Basrah, Iraq

Section 1

Ecological Interpretation of Land Uses Act

Chapter 1

Agriculture and Its Impact on Land-Use, Environment, and Ecosystem Services

Radoslava Kanianska

Additional information is available at the end of the chapter

http://dx.doi.org/10.5772/63719

Abstract

Human expansion throughout the world caused that agriculture is a dominant form of land management globally. Human influence on the land is accelerating because of rapid population growth and increasing food requirements. To stress the interactions between society and the environment, the driving forces (D), pressures (P), states (S), impacts (I), and response (R) (DPSIR) framework approach was used for analyzing and assessing the influence of agriculture on land use, environment, and ecosystem services. The DPSIR model was used to identify a series of core indicators and to establish the nature of interactions between different driving forces, pressures, states, impacts, and responses. We assessed selected indicators at global, national, and local levels. Driving force indicators describe growing population trend and linking land-use patterns. The driving forces exert pressure on the environment assessed by indicators describing development in fertilizer and pesticides consumption, by number of livestock, and by intensification joined growing release of ammonia and greenhouse gas (GHG) emissions from agriculture, and water abstraction. The pressure reflects in the state of environment, mainly expressed by soil and water quality indicators. Negative changes in the state then have negative impacts on landscape, e.g., traditional landscape disappearance, biodiversity, climate, and ecosystem services. As a response, technological, economic, policy, or legislation measures are adopted.

Keywords: Agriculture, land use, environment, ecosystem service, DPSIR model

1. Introduction

Land cover and land-use patterns on Earth reflect the interaction of human activities and the natural environment [1]. Human population growth together with competitive land use causes land scarcity, conversion of wild lands to agriculture and other uses. As we can see, the anthropogenic factor has an important impact on land use and land cover changes. Given this human influence, especially during the past 100 years, the recent period has been called the Anthropocene Age [2]. Human influence on the land and other natural resources is accelerating because of rapid population growth and increasing food requirements. The increasing agricultural intensity generates pressure not only on land resources but also across the whole environment. These factors make agriculture a top-priority sector for both economic and environmental policy.

Comprehensive assessment of the agriculture is a challenging task. There are different possibilities and methods for such assessment. To stress the interactions between society and the environment, the DPSIR framework approach is used for analyzing and assessing the influence of agriculture on land use and environment with emphasis on Slovakia.

2. Methodology

Within integrated environmental assessment a framework is used, which distinguish driving forces (D), pressures (P), states (S), impacts (I), and response (R). This is known as the DPSIR model. As the model can capture the cause–effect relationships between the economic, social, and environmental sectors, it has been widely applied to analyze the interacting processes of human-environmental systems [3]. The DPSIR model originated from the pressure–state–response (PSR) framework, which was developed by the Organisation for Economic Cooperation and Development [4]. Later it was elaborated by European Environment Agency [5]. Environmental indicators should reflect all elements of the chain between human activities, their environmental impacts, and the societal responses to these impacts [6].

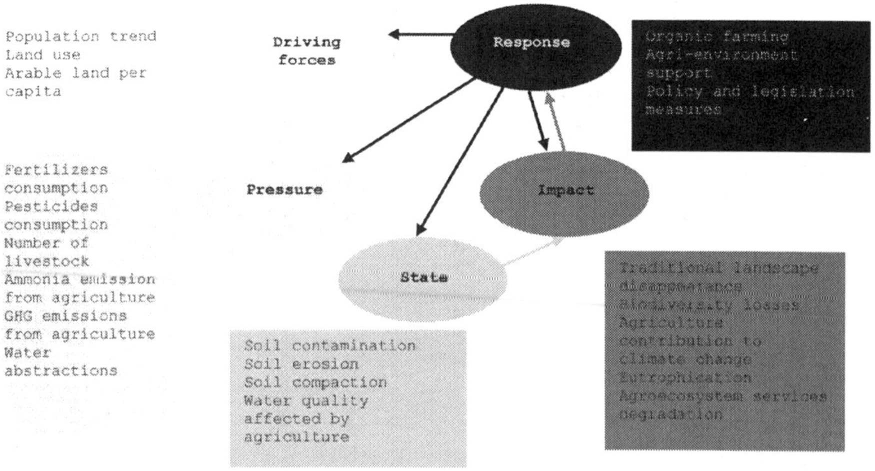

Figure 1. DPSIR model for agriculture and environment.

The DPSIR model was used to identify a series of core indicators and to establish the nature of interactions between the different driving forces, pressures, states, impacts, and responses, and thus to assess the agriculture and its impact on land use, environment, and ecosystem services (**Figure 1**). More attention was paid to Slovakia. We assessed selected indicators at global, national (country Slovakia), and local (cadastre Liptovská Teplička (LT)) level. Slovakia is located in central Europe and covers an area of 49,035 km^2. It is largely located in the mountain territory of the western Carpathian arch. The climate is temperate. Despite the mountain character of the majority of the Slovak territory, there were suitable conditions for agricultural development. The Slovak rural territory represents 87% of the total land area and the Slovak rural population represents 43.7% of the total population. Liptovská Teplička (LT) cadastre is located in the northern part of Slovakia where Low Tatras is adjacent to the Liptov basin with elevation over 900 m above sea level. Mean annual temperature is 5°C, and mean annual precipitation is 900 mm.

3. Results and discussion

3.1. Driving forces

With the growing world population the requirements are grown to cover the food demand. Human expansion throughout the world caused that agriculture is a dominant form of land management globally, and agricultural ecosystems cover nearly 40% of the terrestrial surface of the Earth. Agricultural ecosystems are interlinked with rural areas where more than 3 billion people live, almost half of the world's population. Roughly 2.5 billion of these rural people derive their livelihoods from agriculture. Thus, population and land-use trends are considered to be the main driving forces for agriculture. Besides these driving forces, EEA [7] further distinguished the so-called external and internal driving forces originating from market trends, technological and social changes, as well as the policy framework.

For many economies, especially those of developing countries, agriculture can be an important engine—driving force—of economic growth. Approximately three-quarters of the world's agricultural value added is generated in developing countries where agriculture constitutes the backbone of the economy. But not only in the developing countries but also in the developed countries agriculture has always been the precursor to the rise of industry and services [8].

3.1.1. Population trend

In the twentieth century, the world population grew four times [9]. Although demographic growth rates have been slowing since the late 1970s, the world's population has doubled since then, to approximately 7 billion people currently and is projected to increase to over 9 billion by 2050. But already millions people are still suffering from hunger and malnutrition. The latest available estimates indicate that about 795 million people in the world (just over one in nine) were undernourished in 2014–2016. Since 1990–1992, the number of undernourished people has declined by 216 million globally, a reduction of 21.4%. The vast majority of the hungry people live in the developing regions. The overall hunger reduction trends in the developing countries since 1990–1992 are connected with changes in large populous countries (China, India) [10]. Paradoxically, most of people suffering from hunger and malnutrition are in rural areas and only 20% are in city slums. According to FAO, 50% of them are small peasants, 20% are landless, 10% are nomadic herdsmen or small fishermen, and 20% live in city slums. In the developing countries, this rural social class is, above all, often a victim of marginalization and exclusion from its governing classes (political, economic, and financial) as well as from the urban milieu where there is a concentration of power and knowledge, and therefore money, including funds for development. Often the urban and rural worlds are separated. Whereas in the EU the farming population constitutes only 5% of the total population, it is about 50% in China, 60% in India, and between 60 and 80% in sub-Saharan Africa [11].

In past, Slovakia was typical agrarian country. Even during the nineteenth century the vast majority of the population worked in agriculture, but with the beginning of the twentieth century the decreasing trend began and continued to the present. In 1921, 60.4% of the working population was engaged in agriculture, after 1945, it was 48.1%. In 2012, 50,400 people worked in agriculture [12] which represented 2.2% of the working population, and 2.76 workers worked per 100 ha of agricultural land which was less than EU-27 average (8.81 workers per 100 ha of agricultural land) [13].

3.1.2. Land use

The global land area is 13.2 billion ha. Of this, 12% (1.6 billion ha) is currently in use for cultivation of agricultural crops, 28% (3.7 billion ha) is under forest, and 35% (4.6 billion ha) comprises grasslands and woodland ecosystems. The world's cultivated area has grown by 12% over the past 50 years. Globally, about 0.23 ha of land is cultivated per head of the world's population [14]. In 1960, it was 0.5 ha of cropland per capita worldwide. In Europe, about one-half of land is farmed and arable land is the most common form of agricultural land. Twenty-five percent of Europe's land is covered by arable land and permanent crops, 17% by pastures and mixed mosaics, and 35% by forests. The average amount of cropland and pasture land per capita in 1970 was 0.4 and 0.8 ha and by 2010 this had decreased to 0.2 and 0.5 ha per capita, respectively [15].

Such a state is a result of dynamic land-use and land-cover changes. Humans have altered land cover for centuries, but recent rates of change are higher than ever [16].

Land-use change reflected in land-cover change and land-cover change is a main component of global environmental change [17], affecting climate, biodiversity, and ecosystem services, which in turn affect land-use decision. Land-use change is always caused by multiple interacting factors. The mix of driving forces of land-use change varies in time and space. Highly variable ecosystem conditions driven by climatic variations amplify the pressure arising from high demands on land resources. Economic factors define a range of variables that have a direct impact on the decision making by land managers. Technology can affect labor market and operational processes on land. Demographic factors, such as increase and decrease of popu-

http://dx.doi.org/10.5772/63719

lation, and migration patterns have a large impact on land use. Life-cycle features arise and affect rural as well as urban environments. They shape the trajectory of land-use change, which itself affects the household's economic status.

The development of the present ecosystems in the postglacial period (Holocene) depended on significant changes in climate. Warming in the postglacial period, about 10,000 years ago, created conditions of back migration of individuals species from their refuges, where they were protected during the glacial periods. After the neolitic revolution, human society began to influence more noticeably the development of natural ecosystems. About half of the ice-free land surface has been converted or substantially modified by human activities. Forest covered about 50% of the Earth's land area 8000 years ago, as opposed to 30% today. Agriculture has expanded into forests, savannas, and steppes in all parts of the world to meet the demand for food and fiber.

The central and north Europe were almost completely naturally covered by forests. Only high mountain and alpine rocky localities were without forest cover. Nowadays Europe is a mosaic of landscapes, reflecting the evolutionary pattern of changes that land use has undergone in the past. The greatest concentration of farmland is found in Eastern Europe, where also Slovakia lies, with more than half of its land area in crop cover [18]. Europe is one of the most intensively used continents on the globe. Despite the long tradition of human impact investigation on the environment and vegetation in Europe, there are few comparable studies in North America. This difference is often attributed to the shorter duration of intensive human impact in most of North America versus Europe. As a result, prior studies in the United States have generally been restricted to local investigations [19].

During the past three centuries, in many developing countries and countries with transition economies, growing demand for food due to an increasing population has caused substantial expansion of cropland, accompanied by shrinking primary forests and grassland areas [20]. Based on many studies, in China between 1700 and 1950, cropland area increased and forest coverage decreased. Similarly in India, between 1880 and 2010, cropland area has increased (from 92 to 140.1 million ha), and forest land decreased (from 89 to 63 million ha) [21]. But in the past 50 years, over world rapid urbanization has been evident [22]. Migration in its various forms is the most important demographic factor causing land-use change at timescales of a couple of decades [23]. Rapid economic growth is accompanied by a shift of land from agriculture to industry, infrastructure, road network, and residential use. Countries in East Asia, North America, and Europe have all lost cultivated land during their periods of economic development [18]. The dramatic growth and globalization of China's economy and market since economy reforms in 1978 have brought about a massive loss of croplands, most of which were converted to urban areas and transportation routes during 1978–1995 [24].

In Slovakia land-use trends are in many aspects similar to EU development. In 2013, of the total area of Slovakia agricultural land covered 48.9% (2,397,041 ha) and forest land 41.1% (2,017,105 ha). The highest share of used agricultural land was represented by arable land (58.9%) followed by permanent grasslands (36.1%). The average amount of agricultural land per capita was 0.44 ha [25]. Cereals are the main growing crops. Since 1990, decrease in agricultural land was recorded, often in favor of built-up area. Analysis of historical land-use changes at Liptovská Teplička cadastre showed that the landscape has undergone changes in land-use and cover during the 224 years. From the long-term point of view, gradual afforestation and permanent grassland conversion to forest land was observed where forest land increased from 67.7% in 1782 to 83.7% in 2006 [26].

3.2. Pressure

Agriculture in the last century has evolved from self-sufficiency to surplus in some parts of the world. Thus, transformation was connected with intensification and specialization of production as main trends in European or North American agriculture accompanied by negative impact on the environment. Agricultural intensification is defined as higher levels of inputs and increased output of cultivated or reared products per unit area and time [27]. Over the past 50 years, agricultural production has grown between 2.5 and 3 times, thanks to significant increase in the yield of major crops [14]. Changing land-use practices have enabled world grain harvests to double from 1.2 to 2.5 billion tonnes per year between 1970 and 2010. Globally, since 1970, there has been a 1.4-fold increase in the numbers of cattle and buffalo, sheep and goats, and increases of 1.6- and 3.7-fold for pigs and poultry, respectively [28].

The mix of cropland expansion and agricultural intensification has varied geographically. Tropical Asia increased its food production mainly by increasing fertilizer use and irrigation. Most of Africa and Latin America increased their food production through both agricultural intensification and extensification. In western Africa cropland expansion was accompanied by a decrease in fertilizer use and a slight increase in irrigation [18]. Agriculture is the single largest user of freshwater resources, using a global average of 70% of all surface water supplies.

3.2.1. Intensification and specialization of agriculture

Intensification and specialization have been predominant trends in EU countries including Slovakia for several decades. Between 1965 and 2000 there was a 6.87-fold increase in nitrogen fertilization, a 3.48-fold increase in phosphorous fertilization while irrigated land area expanded 1.68 times, contributing to a 10% net increase in land in cultivation [29]. Strong intensification in Europe in contrast to other countries is obvious if we compare selected indicators, e.g., fertilizer consumption or livestock density (**Figures 2** and **3**). In Slovakia, the maximum intensification level was reached during the socialistic era in 80th. However, since 1990, there are signs of a trend toward a more efficient use of agricultural inputs as a result of not very favorable economic situation of farms but also as a consequence of different environmental measures implementation. During 1980–2010 in Slovakia, indicators concerning to agricultural intensification dropped, in case of fertilizer consumption by 73% (**Figure 4**), the pesticides consumption by 77%. This period is typical in livestock number reduction, in case of cattle by 71, pigs 73, and sheep 37% (**Figure 5**).

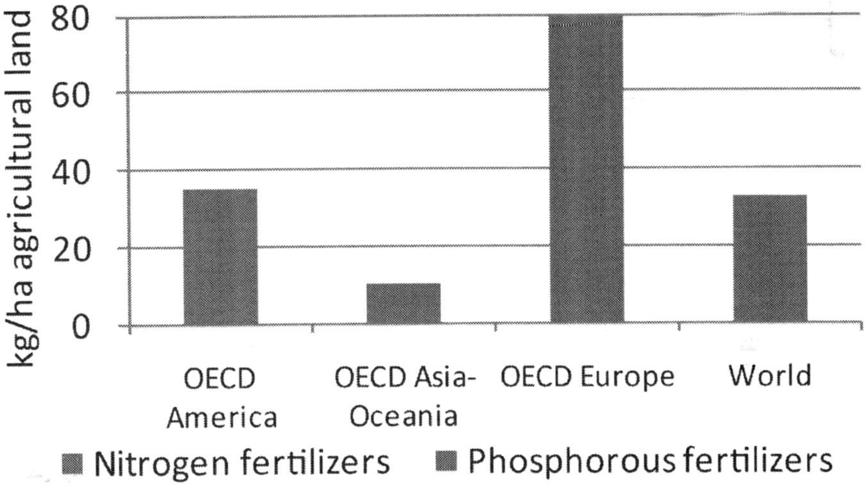

Figure 2. Fertilizer consumption in 2012 (kg/ha of agricultural land) (based on data from OECD [30]).

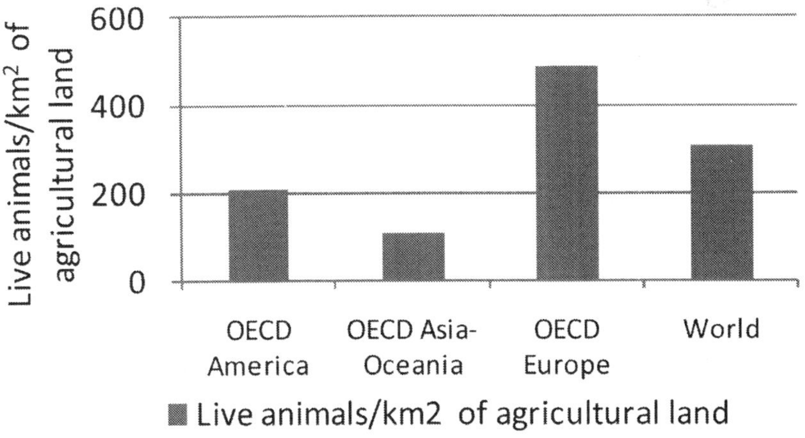

Figure 3. Livestock density in 2012 (live animals/km^2 of agricultural land) (based on data from OECD [30]).

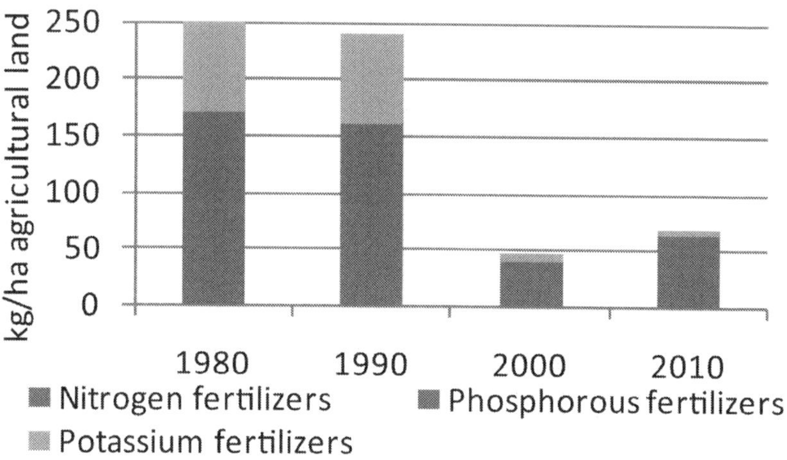

Figure 4. Development in fertilizer consumption in Slovakia (kg pure nutrient/ha) (based on data from CCTIA [31]).

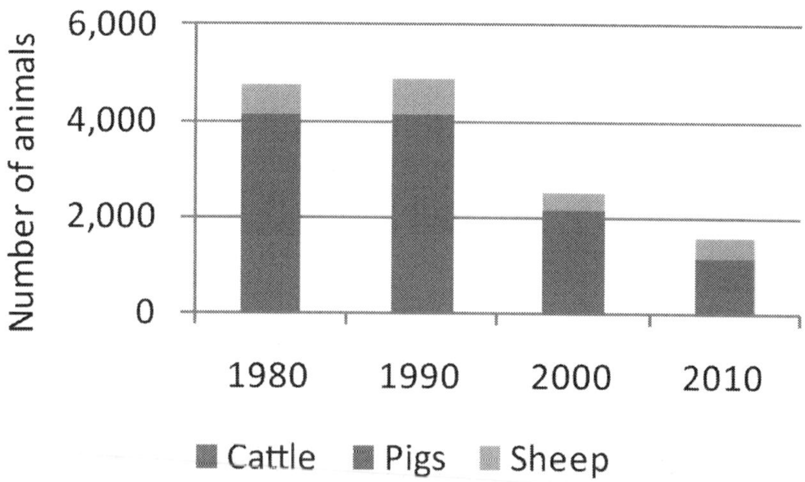

Figure 5. Development in number of livestock in Slovakia (live animals/ha of agricultural land) (based on data from SOSR [32]).

Intensification is connected with increasing release of atmospheric emissions through management of land and livestock, and thus agriculture release to the atmosphere significant amounts of greenhouse gases emissions of CO_2, CH_4, and N_2O [33] and ammonia emissions. The agricultural sector is currently responsible for the vast majority of ammonia emissions in

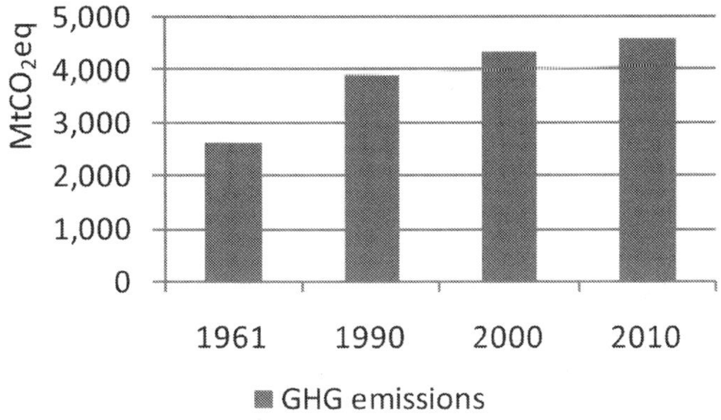

Figure 6. Global GHG annual agriculture emissions (MtCO2eq) (based on data from Tubiello et al. [35]).

the European Union. Agriculture contributes to about 47 and 58% of total anthropogenic emissions of CH_4 and N_2O, respectively. Annual GHG emissions from agricultural production in 2000–2010 were estimated at 5.0–5.8 $GtCO_2eq$/year while annual GHG flux from land use and land-use change activities accounted for approximately 4.3–5.5 $GtCO_2eq$/year. The enteric fermentation and agricultural soils represent together about 70% of total emissions, followed by paddy rice cultivation (9–11%), biomass burning (6–12%), and manure management (7–8%) [34]. Development of the global GHG annual agriculture emissions from 1961 to 2010 based on FAOSTAT data shows **Figure 6**. Annual GHG emissions from agriculture are expected to increase in coming decades due to escalating demands for food and shift in diet. However improved management practices and emerging technologies may permit a reduction in emissions per unit of food produced. In Slovakia, due to decrease number of livestock also decreasing trend in GHG and ammonia emissions were observed since 1990 (**Figure 7**).

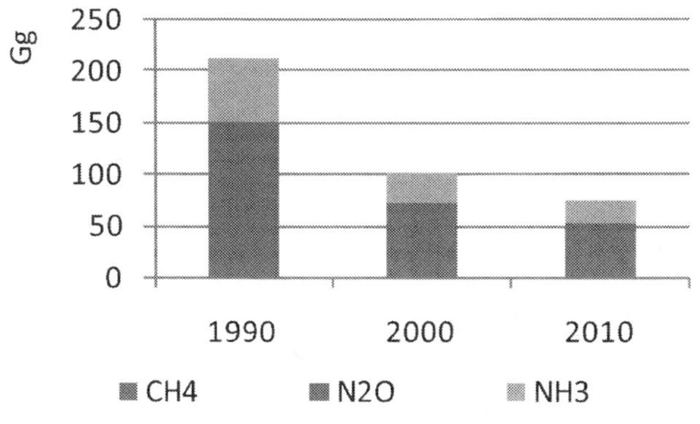

Figure 7. Emissions from agriculture in Slovakia (Gg) (based on data from MESR, SEA [36]).

3.3. State

Intensive management practices in agriculture escalating rates of land degradation threatens most crop and pasture land throughout the world. Worldwide, more than 12 million hectares of productive arable land are severely degraded and abandoned annually. Increased pressure is connected with deterioration of the state of environment, mainly soil and water.

3.3.1. Soil

Soil is the most fundamental asset on farms. Its quality that directly affects provisioning ecosystem services is strongly affected by management practices. The state of soils can be assessed by the help of indicators on soil contamination, erosion, and compaction.

Soil contamination implies that the concentration of a substance in soil is higher than would naturally occur. Agricultural activities contribute to soil contamination by introducing pollutants or toxic substances such as cadmium by application of mineral phosphate fertilizers or organic pollutants by pesticide application. Comprehensive inventories and databases on local and diffuse soil contamination are lacking on the global or regional extent. Estimates show that about 15% of land in the EU-27 exhibits a surplus in excess of 40 kg N/ha [37]. In Slovakia, data from the soil monitoring showed that only 0.4% of the total soil cover is contaminated by heavy metals [38].

The loss of soil from land surfaces by soil erosion has been significantly increased by human activities. Each year about 10 million ha of cropland are lost due to soil erosion [39]. In Slovakia, 32% of agricultural land is threatened by water and 5% by wind erosion, respectively [36].

Since the 1950s, pressure on agricultural land has increased considerably also owing to agricultural modernization and mechanization what caused next serious environmental problem—soil compaction. Overuse of machinery, intensive cropping, short crop rotations, intensive grazing, and inappropriate soil management leads to compaction [40]. Soil compaction problems, in various degrees, are found in virtually all cropping systems throughout the world. They are of particular significance where intensive mechanization has been adopted on soils subject to high rainfall or irrigation [41]. According to estimation approximately 600,000 ha of agricultural land is compacted in Slovakia [42].

The effect of farming on soil causing soil compaction expressed as soil penetrometric resistance (PR measured to 20 cm depth in MPa) was investigated in May 2014 at Liptovská Teplička cadastre, on soil type Rendzina with four different land use (AL, arable land; M, meadow; AG, abandoned grasslands; FL, forest land) (**Figure 8a–d**). The different land use and practices reflected in different PR values (**Figure 9a–d**). The highest mean PR value was measured in AL (1.52 MPa), followed by M and FL (same value of 1.08 MPa), and abandoned grasslands (0.90 MPa) [43]. Measured values show at compaction in arable land. But there is necessary to take into account possibility that PR value in AL could be also the lowest among observed different land-use sites. Such situation can be observed when the measurement is done immediately after some technological operation, e.g., ploughing, contributing to turning the soil over, and diminishing higher soil horizons compaction.

Figure 8. (a) Arable land in cadastre Liptovská Teplička, Law Tatras Mountain.

Figure 8. (b) Meadow in cadastre Liptovská Teplička, Law Tatras Mountain.

Figure 8. (c) Abandoned grasslands in cadastre Liptovská Teplička, Law Tatras Mountain.

Figure 8. (d) Forest land in cadastre Liptovská Teplička, Law Tatras Mountain.

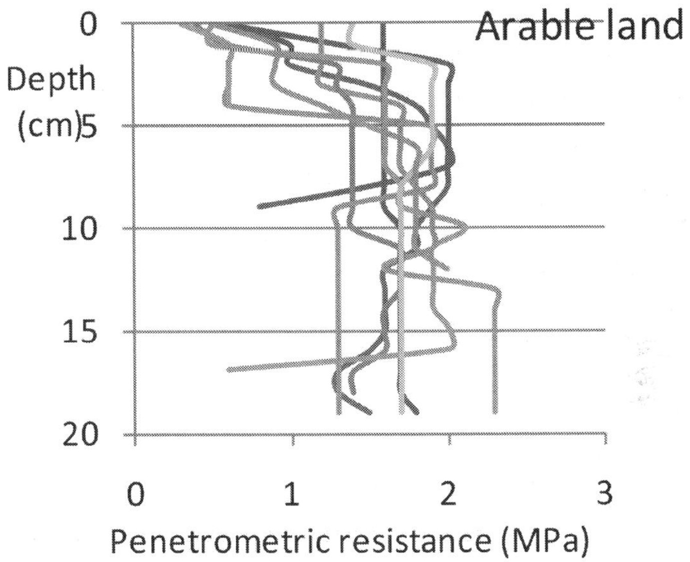

Figure 9. (a) Penetrometric resistance at arable land in cadastre Liptovská Teplička.

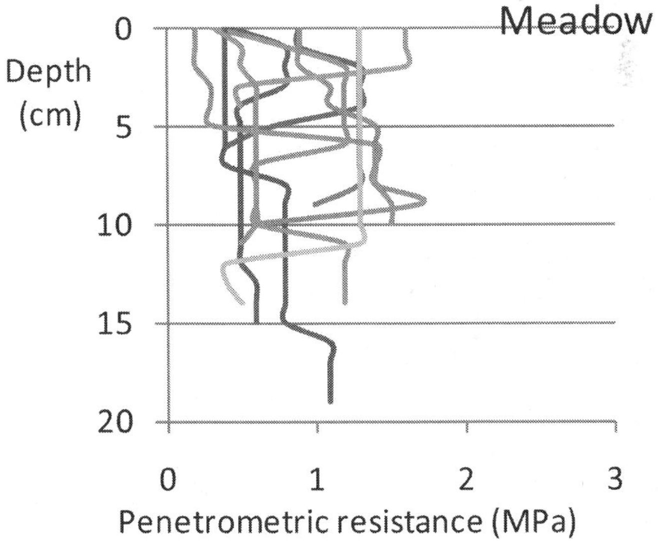

Figure 9. (b) Penetrometric resistance at meadow in cadastre Liptovská Teplička.

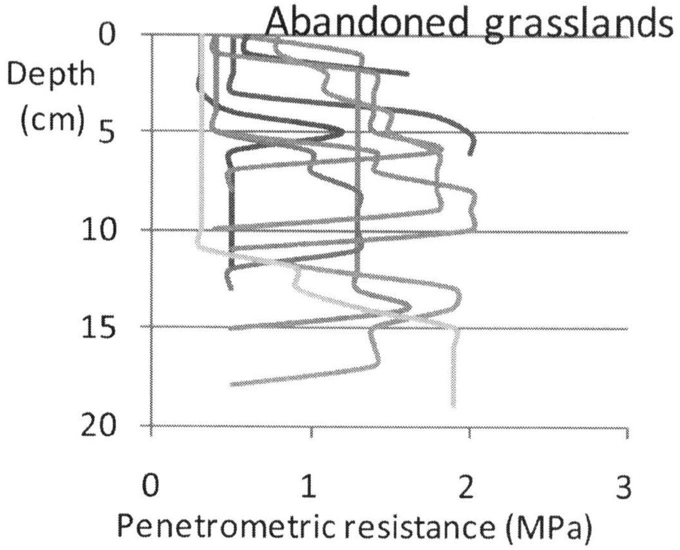

Figure 9. (c) Penetrometric resistance at abandoned grasslands in cadastre Liptovská Teplička.

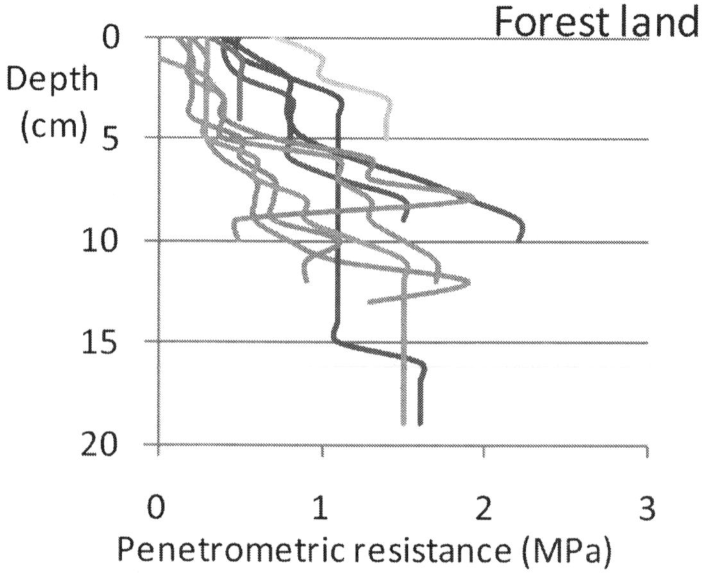

Figure 9. (d) Penetrometric resistance at forest land in cadastre Liptovská Teplička.

3.3.2. Water

Agriculture is both cause and victim of water pollution. Evidence for elevated nitrate and phosphate contents on farm, in drains, streams and rivers, and lakes is partial and tends to be specific to a given location and circumstance. Global phosphorus flux to the ocean increased 3-fold to about 22 Tg per year by the end of the twentieth century.

Nitrate is the most common chemical contaminant in the world's aquifers. An estimate for continental USA in the 1990s indicates that returns to water are close to 20% of total applied agricultural nitrogen, with up to 25% lost in gaseous form. Mean nitrate levels have increased by about 36% in global waterways since 1990 [44].

Pesticides contaminate surface water and groundwater. They can reach surface water through runoff from treated plants and soil. Contamination of water by pesticides is widespread, and groundwater pollution due to pesticides is a worldwide problem [45].

3.4. Impact

Impacts are commonly the result of multiple stressors. Agriculture exerts pressure on the environment that is both beneficial and harmful and can result in both positive and negative environmental impacts. The wide variation in farming systems and practices throughout the world, and differing environmental characteristics mean that the effects of agriculture on the environment arise at site-specific level but can have impacts at local to global level.

3.4.1. Traditional landscape disappearance

The disappearance of traditional agricultural landscape is an ongoing process, accompanying the general trend of agricultural abandonment in Europe [46]. In Slovakia, traditional agricultural landscape is described as agricultural ecosystems that consist of mosaics of small-scale arable fields or permanents agricultural cultivations such as grasslands, vineyards, and high-trunk orchards or early abandoned plots with a low succession degree [47]. Important parts of such landscape are linear landscape elements (hedges, tree lines, stone walls).

In Slovakia, traditional extensive farming with individual farmer attitude to landscape was transformed to collectivization with overall interest in land exploitation [48]. Collectivization caused small-scale parcels managed by individual farmers to be consolidated into large blocks (polygons) managed by large co-operative farms and resulted in a decrease of the mosaic of arable land and grasslands. At Liptovská Teplička cadastre during 1956–1990, number of polygons decreased from 15 to 2 at arable land, and from 82 to 29 at permanent grasslands [26]. In addition, the management of traditional agricultural landscapes structures decreased rapidly after collectivization. Nowadays the main barriers in ideal management are unfavorable subsidies in agriculture and the financial inaccessibility of modern tools and machinery together with inadequate market and the weak support of local government [49].

3.4.2. *Contribution to climate change*

Anthropogenic land-use activities and changes in land use/cover caused changes superimposed on the natural fluxes. Land-cover changes are responsible for surface and vegetation modifications what reflects in surface albedo and thus surface-atmosphere energy exchanges, which have an impact on regional climate. Terrestrial ecosystems are important sources and sinks of carbon and thus land-use changes reflect also in the carbon cycle. The important contribution of local evapotranspiration to the water cycle—that is precipitation recycling—as a function of land cover highlighted yet another considerable impact of land-use/cover change on climate, at a local to regional scale [50].

The influence of land use/cover on soil temperature was investigated at Liptovská Teplička cadastre study site in May 2014 where 10 measurements in depth of 5 and 25 cm at four different land-use plots (AL, arable land; M, meadow; AG, abandoned grasslands; FL, forest land) were done by insert soil thermometer (**Table 1**). The highest mean soil temperature was recorded in AL in 5 cm depth (4.6°C), the lowest in FL in 5 cm depth (3.5°C). Measured values show how plant cover and its microclimate functions are important and can affect soil temperature.

Depth (cm)	**Land use**			
	Arable land	**Meadow**	**Abandoned grasslands**	**Forest land**
5	4.6	4.3	4.2	3.5
25	4.3	4.4	4.6	3.8

Table 1. Actual soil temperature in cadastre Liptovská Teplička in May 2014 (°C).

Agriculture is unique among economic sectors releasing GHG emissions and thus contributing to climate change. Agricultural activities lead, in fact, not only to sources but also to important sinks of CO_2. Agricultural contribution to greenhouse gases accounts for 13.5% of global greenhouse gas emissions [51]. At the same time, agricultural production is fully climate and several further natural conditions dependent. Every change in climate has not only short-term but also long-term consequences. Climate change brings an increase in risk and unpredictability for farmers—from warming and related aridity, from shifts in rainfall patterns, and from the growing incidence of extreme weather events.

On the other hand, agriculture can also positively contribute to climate change mitigation. The utilization of agricultural residues as raw materials in a biorefinery is a promising alternative to fossil resources for production of energy carriers and chemicals, thus mitigating climate change and enhancing energy security [52].

3.4.3. *Biodiversity losses*

Land use, specifically in agriculture, has great impact on biodiversity. Another aspect contributing to biodiversity decline is that humans today depend for survival on tiny fraction of wild species that has been domesticated. Yet only 14 of 148 species weighing 45 kg or more were actually domesticated. Similarly, worldwide there are about 200,000 wild species of higher plants, of which only about 100 yielded valuable domesticates [53].

All long-term historical land-use changes responsible for natural ecosystems conversion to seminatural ecosystems or artificial systems contributed to the extensive changes in biodiversity composition and ecological processes. Agriculture plays an important role in these processes and is responsible for biodiversity decline. Over the past 50 years, ecosystems have changed more rapidly than at any other period of human history [62]. This period is connected with high agricultural intensification in many parts of the world. Land-use changes have been shown to be one of the leading causes of biodiversity loss in terrestrial ecosystems [54, 55]. To demonstrate the impact of land use and land management on soil biota quantitative analysis of earthworm was done at Liptovská Teplička cadastre in May 2014 when earthworms were hand sorted, weighted, and numbered from seven soil monoliths (35 cm × 35 cm × 20 cm) placed in line in 3 m distance in four different land-use plots (AL, arable land; M, meadow; AG, abandoned grasslands; FL, forest land). The earthworms may be used as bioindicator because they are very sensitive to both chemical and physical soil parameters. Earthworm biomass or abundance can offer a valuable tool to assess different environmental impacts such as tillage operations, soil pollution, different agricultural input, trampling, and industrial plant pollution [56]. The highest mean number (87.5 individuals m^{-2}) and earthworm body biomass (40.3 g m^{-2}) was recorded in M, the lowest in AG (5.8 individuals m^{-2} and 5.9 g m^{-2} body biomass) (**Table 2**) [49]. Relatively high number and earthworm biomass in AL at Liptovská Teplička cadastre is consequence of organic farming.

Depth (cm)	Land use			
	Arable land	**Meadow**	**Abandoned grasslands**	**Forest land**
Number	33.8	87.5	5.8	8.2
Body biomass	16.2	40.3	5.9	6.6

Table 2. Number of earthworm individuals and earthworm body biomass in cadastre Liptovská Teplička in May 2014 (individuals m^{-2}, g m^{-2}) [43]

Though intensified land use is undeniably the main cause of biodiversity loss. There is an increasing expectation that productive agricultural landscapes should be managed to preserve or enhance biodiversity [57].

3.4.4. *Eutrophication*

Eutrophication is a process of pollution that occurs when a lake or stream becomes overrich in plant nutrients as a consequence it becomes overgrown in algae and other aquatic plants. The major impacts of eutrophication due to overloading with nitrogen and phosphorus nutrients are changes in the structure and functioning of marine ecosystems, reduced biodiversity, and reduced income from fishery, mariculture, and tourism. The main source of nitrogen run-off from agricultural land brought to the sea via rivers. Atmospheric deposition of nitrogen may also contribute significantly to the nitrogen load. This nitrogen originates partly from ammonia evaporation from animal husbandry. Most of the phosphorus comes from households and industries discharging treated or untreated wastewater to freshwater directly to the sea, and from soil erosion.

Human activity has increased N fluxes. In 1970s, an explosive increase in coastal eutrophication in many parts of the world correlates well with the increased production of reactive N for agriculture and industry [45]. Eutrophication is a global environmental problem. In EU, there is marked variation in groundwater nitrate concentration between different geographical regions with high concentration in Western Europe and very low concentrations in Northern Europe. The lack of a general decrease is due to continued high emissions from agriculture [58].

3.4.5. Agroecosystem services degradation

Agroecosystems both provide and rely on ecosystem services to sustain production food, fiber, and other harvestable goods. Increases in food and fiber production have often been achieved at the cost of other critical services.

Services that help to support production of harvestable goods can be considered as services to agriculture. These services include soil structure and fertility enhancement, nutrient cycling, water provision, erosion control, pollination, and pest control, among others. Ecological processes that detract from agricultural production can be considered disservices to agriculture and include pest damage, competition for water, and competition for pollination. Management of agricultural ecosystems also affects flows of ecosystem services and disservices (or diminution of naturally occurring services) from production landscape to surrounding areas. Disservices from agriculture can include degradation or loss of habitat, soil, water quality, and other off-site, negative impacts [59].

Provision of ecosystem services in farmlands is directly determined by their design and management [60] and strongly influenced by the function and diversity of the surrounding landscape [61]. The Millenium Ecosystem Assessment [62] reported that approximately 60% (15 out of 24) of services measured in the assessment were being degraded or unsustainably used as a consequence of agricultural management and other human activities.

3.5. Response

In recent decades, increasing concern for the environment and sustainability has compelled many governments to continuously adjust their land-use policies to balance multiple uses of land resources. These policies have caused changes in cropland and its spatial distribution. There are different environmental objectives incorporated into agrienvironment measures, training programs, support for investments in agricultural holdings, protection of the environment in connection with agriculture and landscape conservation, support to improving the processing and marketing of agricultural products. Organic farming or low-input farming systems are examples where support for the processing or marketing of their products can help in achieving environmental objectives. In 2013, there were 43.1 million hectares of organic agricultural land, including conversion areas. The regions with the largest areas of organic agricultural land are Oceania and Europe [63]. In Slovakia, organic farming area covered 8.4% of the total agricultural land [36].

4. Conclusion

Agriculture is a dominant form of land management globally. Rapid population growth as primary driving force connected with increasing food requirements generate great pressure on future land use, environment, natural resources, and ecosystem services. The DPSIR framework approach helped us to analyze selected indicators having the cause–effect relationships between the economic, social, and environmental sectors.

Recent rates of land-use and cover changes are higher than ever. In many developing countries and countries with transition economies, growing demand for food has caused expansion of cropland. Extensive agricultural systems are slowly intensified. In developed countries, economic growth has been recently accompanied by a shift of land from agriculture to industry, road network, and residential use. Extensive forms of agriculture used in past mainly in Europe and North America were transformed into industrial-style agriculture accompanied by intensification and specialization. The large inputs of fertilizers, pesticides, fossil fuels have large, complex effects on the environment. Agriculture releases significant amounts of greenhouse gases and ammonia emission to the atmosphere. It is the single largest user of freshwater resources. Intensive management practices escalating rates of land degradation, soil and water deterioration. The effects on the environment arise at site-specific level but can have impact at local to global levels. Land-cover changes cause the disappearance of traditional agricultural landscape and are responsible for vegetation modifications which have an impact on regional climate, carbon sequestration, and biodiversity losses. Agriculture also has impact on the natural systems and ecosystem services on which humans depend.

Future challenges relating to greater pressure on environment, natural resources, and climate change imply that a "business as usual" model in agriculture is not a viable option. Green growth is a new method that places strong emphasis on the complementarities between the economic, social, and environmental dimensions of sustainable development. Thus, the main role of future agriculture is its transformation into good productive but a sustainable system that can be effective for centuries without adverse effect on natural resources on which agricultural productivity depends.

Acknowledgements

This work was supported by the Slovak Research and Development Agency under Grant No. APVV-0098-12 Analysis, modeling and evaluation of agro-ecosystem services. The research of abiotic soil parameters was done by the equipment supported by Operational Programme Research and Development via contract No. ITMS-26210120024 Restoration and building of infrastructure for ecological and environmental research at Matej Bel University in Banská Bystrica.

Author details

Radoslava Kanianska

Address all correspondence to: radoslava.kanianska@umb.sk

Faculty of Natural Sciences, Matej Bel University, Banská Bystrica, Slovakia

References

[1] Alonso-Pérez F, Ruiz-Luna A, Turner J, Berlanga-Robles CA, Mitchelson-Jacob G. Land cover changes and impact of shrimp aquaculture on the landscape in the Ceuta coastal lagoon system, Sinaloa, Mexico. Ocean & Coastal Management. 2003;46(6–7):583–600.

[2] Slaughter RA. Welcome to the anthropocene. Futures. 2011;44(2):19–26.

[3] Pinto R, de Jonge VN, Neto JM, Domingos T, Marques JC, Patrício J. Towards a DPSIR driven integration of ecological value, water uses and ecosystem services for extuarine systems. Ocean & Coastal Management. 2013;72:64–79.

[4] OECD. OECD core set of indicators for environmental performance reviews. OECD Environmental Directorate Monographs no. 83. Paris: OECD; 1993. 39 p.

[5] Burkhard B, Müller F. Driver–pressure–state–impact–response. In: Jorgensen SE, Fath BD, editors. Ecological indicators. Vol. 2 of Encyclopedia of ecology. Oxford: Elsevier; 2008. p. 967–970.

[6] Gabrielsen P, Bosch P. Environmental indicators: typology and use in reporting. Copenhagen: EEA; 2003. 19 p.

[7] EEA. Integration of environment into EU agriculture policy – the IRENA indicator-based assessment report. Copenhagen: EEA; 2006. 64 p.

[8] FAO. FAO statistical yearbook 2013. World food and agriculture. Rome: FAO; 2013. 307 p.

[9] UNEP. Towards a green economy: pathway to sustainable development and poverty reduction. A synthesis for policy makers. Nairobi: UNEP; 2011. 52 p.

[10] FAO, IFAD, WFP. The state of food insecurity in the world 2015. Meeting the 2015 international hunger targets: taking stock of uneven progress. Rome: FAO; 2015. 62 p.

[11] Feyder J. Commentary I: agriculture: a unique sector in economic ecological and social terms. In: Trade and environment review 2013. Wake up before it is too late. Make

agriculture truly sustainable now for food security in a changing climate. Geneva: UNCTAD; 2013. p. 9–12.

[12] MARDSR. Report on agriculture and food industry in the Slovak republic. Green report. Bratislava: MARDSR; 2013. 68 p. (in Slovak).

[13] Szabo L, Grznár M. Labour and performance of agriculture in the Slovak republic. Economics of Agriculture. 2015;XV/3:4–13 (in Slovak).

[14] FAO. The state of the world's land and water resources for food and agriculture (SOLAW) – managing systems at risk. Rome: FAO and London: Earthscan; 2011. 308 p.

[15] EEA. The European environment – state and outlook 2010: synthesis. Copenhagen: EEA; 2010. 212 p.

[16] Hansen MC, Stehman SV, Potapov PV. Quantification of global gross forest cover loss. Proceedings of the National Academy of Sciences of the United States America. 2010;107:8650–8655.

[17] Foley JA, DeFries RS, Asner GP, Barford C, Bonana G, Carpenter SR, Chapin FS, et al. Global consequences of land use. Science. 2005;80(309):570–574.

[18] Ramankutty N, Foley JA, Olejniczak NJ. People on the land: changes in global population and croplands during the 20th Century. AMBIO: A Journal of the Human Environment. 2002;31:251–257.

[19] Foster DR. Land-use history (1730–1990) and vegetation dynamics in central New England, USA. Journal of Ecology. 1992;80:753–772.

[20] Liu M, Tian H. China's land cover and land use change from 1700 to 2005: estimations from high-resolution satellite data and historical archives. Global Biogeochemical Cycles. 2010;24:1–18.

[21] Tian H, Banger K, Bo T, Dadhwal VK. History of land use in India during 1880–2010: large-scale land transformations reconstructed from satellite data and historical archives. Global and Planetary Change. 2014;121:78–88.

[22] Miao L, Zhu F, He B, Ferrat M, Liu Q, Cao X, Cui X. Synthesis of China's land use in the past 300 years. Global and Planetary Change. 2013;100:224–233.

[23] Geist HJ, Lambin EF. Proximate causes and underlying driving forces of tropical deforestation. BioScience. 2002;52(2):143–50.

[24] Chen J. Rapid urbanization in China: a real challenge to soil protection and food security. Catena. 2007;69:1–15.

[25] IGCCSR. Statistical yearbook on land resources in the Slovak republic. Bratislava: IGCCSR; 2015. 130 p. (in Slovak).

[26] Kanianska R, Kizeková M, Nováček M, Zeman M. Land-use and land-cover changes in rural areas during different political systems: a case study of Slovakia from 1782 to 2006. Land Use Policy. 2014;36:554–566.

[27] Matson PA, Parton WJ, Power AG, Swift MJ. Agricultural intensification and ecosystem properties. Science. 1997;277:504–509.

[28] FAOSTAT. FAOSTAT database. Food and Agriculture Organisation of the United Nations; 2013. Available at: http://faostat.fao.org/.

[29] Tilman D, Socolow R, Foley JA, Hill J, Larson E, Lyind L, Pacala S, Reilly J, Searchinger T, Somerville C, Williams R. Beneficial biofuels – the food, energy, and environment trilemma. Science. 2009;325:270–271.

[30] OECD. Environment at a glance 2015: OECD indicators. Paris: OECD; 2015. 104 p.

[31] CCTIA. Results of agrochemical soil testing in Slovakia during 2006–2011. XII period. Bratislava: CCTIA; 2013. 96 p. (in Slovak).

[32] SOSR. Inventory of livestock (to 30.11.2014). Bratislava: SOSR; 2015. 23 p. (in Slovak).

[33] Paustin K, Babcock BA, Hatfield J, Lal R, McCarl BA, McLaughhlin S, Mosier A. et al. Agricultural mitigation of greenhouse gases: science and policy options. CAST report; 2004. 18 p.

[34] Smith PM, Bustamante H, Ahammad H, Clark H, Dong EA, Elsiddig H, Haberl R. et al. Agriculture, forestry and other land use (AFOLU). In: Climate change 2014: mitigation of climate change. Contribution of working group III to the fifth assessment report of the IPCC. Cambridge, UK and USA: Cambridge University Press; 2014. 112 p.

[35] Tubiello FN, Salvatore M, Rossi S, Ferrara A, Fitton N, Smith P. The FAOSTAT database of greenhouse gas emissions from agriculture. Environmental Research Letters. 2013;8:1–10.

[36] MESR, SEA. State of the environment report of the Slovak republic 2014. Bratislava, Banská Bystrica: SEA; 2015. 208 p.

[37] JRC IES EC. The state of soil in Europe. A contribution of the JRC to the EEA's environment state and outlook report-SOER 2010. Ispra: JRC IES EC; 2012. 78 p.

[38] Kobza J. Soil and plant pollution by potentially toxic elements in Slovakia. Plant, Soil and Environment. 2005;51:243–248.

[39] Pimentel D, Burgess M. Soil erosion threatens food production. Agriculture. 2013;3:443–463.

[40] Hamza MA, Anderson WK. Soil compaction in cropping systems. A review of the nature, causes and possible solutions. Soil & Tillage Research. 2004;82:121–145.

[41] Soane BD, Ouwerker C. Soil compaction in crop production. Developments in agricultural engineering 11. Netherlands: Elsevier; 1994. 684 p.

[42] Fulajtár E. Assessment and determination of the compacted soils in Slovakia. In: Advanced in geoecology. Catena Verlag; 2000. p. 384–387.

[43] Kanianska R, Jadudová J. Evaluation of selected biotic and abiotic soil parameters having impact on ecosystem services. In: Kukla J, Kuklová M, editors. Proceedings, Zvolen 11 June 2015. Bratislava: SSPLPVV SAV, Zvolen: ÚEL SAV; 2015. p. 32–36.

[44] WWDR4. Managing water along the livestock value chain. Chapter 18. World water development report. Rome: FAO; 2011.

[45] Turral H, Mateo-Sagasta X, Burke J. Water pollution from agriculture: a review. Rome: FAO; 2012. 173 p.

[46] Gerard F, Petit S, Smith G, Thomson A, Brown N, Manchester S, Wadsworth R. et al. Land cover change in Europe between 1950 and 2000 determined employing aerial photography. Progress in Physical Geography. 2010;34:183–205.

[47] Dobrovodská M, Špulerová J, Štefunková D, Halabuk A. Research and maintenance of biodiversity in historical structures in the agricultural landscape of Slovakia. In: Barančoková M, Krajči J, Kollár J, Belčáková I, editors. Landscape Ecology – Methods, Applications and Interdisciplinary Approach. Bratislava: ILE SAS; 2010. p. 131–140.

[48] Bezák P, Petrovič F. Agriculture, landscape, biodiversity: scenarios and stakeholder perceptions in the Poloniny national park (NE Slovakia). Ecology. 2006;25(1):82–93.

[49] Lieskovský J, Bezák P, Špulerová J, Lieskovský T, Koleda P, Dobrovodská M, Bürgi M, Gimmi U. The abandonment of traditional agricultural landscape in Slovakia – analysis of extent and driving forces. Journal of Rural Studies. 2015;37:75–84.

[50] Lambin EF, Geist HJ, Lepers E. Dynamics of land-use and land-cover change in tropical regions. Annual Reviews of Environmental Resources. 2003;28:205–241.

[51] IPCC. The fourth assessment report of the Intergovernmental Panel on climate change. Geneva: IPCC; 2007. 112 p.

[52] Cherubini F, Ulgiati S. Crop residues as raw materials for biorefinery systems – a LCA case study. Applied Energy. 2010;87:47–57.

[53] Diamond J. Evolution, consequences and future of plant and animal domestication. Nature. 2002;418:700–707.

[54] Daily GC, Polasky S, Goldstein J, Kareiva PM, Mooney HA, Pejchar L. et al. Ecosystem services in decision making: time to deliver. Frontiers in Ecology and the Environment. 2009;7(1):21–28.

[55] Reidsma P, Telenburg T, van den Berg M, Alkemade R. Impacts of land-use change on biodiversity: an assessment of agricultural biodiversity in the European Union. Agriculture, Ecosystems and Environment. 2006;114(1):86–102.

[56] Paoletti MG. The role of earthworms for assessment of sustainability and as bioindicators. Agriculture, Ecosystems and Environment. 1999;74:137–155.

[57] Weeks ES, Mason N. Prioritising land-use decisions for the optimal delivery of ecosystem services and biodiversity protection in productive landscape. In: Grillo O, editor. Biodiversity – the dynamic balance of the planet. Rijeka, Croatia: InTech; 2014. p. 1–32.

[58] EEA. Eutrophication in Europe's coastal waters. Topic report 7/2001. Copenhagen: EEA; 2001. 86 p.

[59] Garbach K, Milder JC, Montenegro M, Karp DS, DeClerck FAJ. Biodiversity and ecosystem services in agroecosystem. In: Van Alfen N, editor. Encyclopedia of Agriculture and Food Systems, Volume 2. Netherlands: Elsevier; 2014. p. 21–40.

[60] Zhang W, Ricketts T, Kremen C, Carney K, Swinton S. Ecosystem services and dis-services to agriculture. Ecological Economics. 2007;64:253–260.

[61] Kremen C, Ostfeld R. A call to ecologists: measuring, analysing, and managing ecosystem services. Frontiers in Ecology and the Environment. 2005;3:540–548.

[62] MEA. Millenium ecosystem assessment synthesis report. USA, Washington D.C.: Island Press; 2005. 155 p.

[63] Willer H, Lernoud J. The world of organic agriculture 2015: summary. In: RIOA FiBL & IFOAM: the world of organic agriculture. Statistics and emerging trends; 2015. p. 24–30.

Chapter 2

Modelling the Contribution of Land Use to Nitrate Yield from a Rural Catchment

Maria-Luz Rodríguez-Blanco, Ricardo Arias,
Maria-Mercedes Taboada-Castro,
Joao Pedro Nunes, Jan Jacob Keizer and
Maria-Teresa Taboada-Castro

Additional information is available at the end of the chapter

http://dx.doi.org/10.5772/63718

Abstract

The nutrient flow dynamics in rural landscapes are among the basic characteristics of landscape functioning. In this study, the ecohydrological model SWAT (Soil and Water Assessment Tool) was applied in a small rural catchment in northwest (NW) Spain to evaluate the contribution of land use on nitrate losses and to assess the relative importance of different pathways by which nitrate is delivered to the drainage network. The model was first calibrated and validated at a monthly time step. The SWAT model performance was satisfactory ($R^2 > 0.5$; Nash-Sutcliffe efficiency (NSE) > 0.5 and percent bias (PBIAS) < 10%) during both the calibration and validation periods, indicating that SWAT predicted the nitrate discharge accurately. Using the calibrated SWAT model, this study showed that agricultural lands, even though they represent only 30% of the catchment, were main contributor to the nitrate losses accounting for about 77% of the total nitrate yield. The model results also indicated that, irrespective of the land use, groundwater flow is the main pathway for nitrate losses (63%); therefore, appropriate management practices aimed at decreasing nitrate leaching will be key factors in reducing nitrate yield in the study catchment.

Keywords: nitrate yield, rural landscape, modelling approach, land use, NW Spain

1. Introduction

The nutrient flow dynamics in rural landscapes are among the basic characteristics of landscape functioning [1]. Nutrient cycling has been well documented in this type of catchment, and a great deal of research has focused on analysing the impact of human activity on nutrient losses [2–4], identifying agricultural activities as the primary source of diffuse pollution of water resources [4, 5]. One of the major diffuse pollutants that is sourced from agriculture is nitrate (NO_3-N), which, due to its high solubility, is easily leached from the soil to both ground and surface waters. In fact, nitrogen leaching from agricultural land has become a common problem in many European regions [6]. An example is Galicia (NW Spain), which has been identified as a European region of high soil and water eutrophication risk [7]. In fact, in less than 10 years, an increase in nitrate concentrations from 2–3 to 10–20 mg l^{-1} was detected in many rural areas [2].

The Water Framework Directive (WFD) demands the implementation of measures in order to improve water quality, the ultimate WFD target being that all waters in the European Union should be in good ecological condition by the end of 2015 [8]. This requires knowledge of the effects of natural conditions and land uses on diffuse pollutant losses at catchment scale and to understand the pathways transporting these pollutant losses from land to water bodies. Different approaches have been adopted to address these issues and the control of diffuse pollutants at source (e.g., through efficient land management practices) is often seen as the optimal solution to potential problems. However, conducting field experiments to better understand diffuse-source pollution and design appropriate management solutions is expensive, time consuming and spatially impractical at catchment scale [9]. In this context, simulation models have become useful tools to evaluate water quality under current conditions and investigate the consequences of land use, management and climate change on water quality. Therefore, they would be helpful to find appropriate measures for assessing environmental and ecological status, taking into account factors such as climate, land and water use, with these becoming vital tools in catchment management. However, before any process-based catchment model can be applied, the performance and reliability of the model must be tested with measured data otherwise model simulations may lead to erroneous results and to faulty design of protection measures. A large number of hydrological models, such as AGNPS (Agricultural Non-point Source Pollution Model), ANSWER (Areal Non-point Source Watershed Environment Response Simulation) and SWAT (Soil and Water Assessment Tool), have been applied to assess these issues. The SWAT [10, 11] is one of the few physically based approaches describing processes responsible for the transfer of nutrients from soil to water with an explicit representation of plant growth and impact of agricultural management practices. It has been a widely used and scientifically accepted tool for modelling diffuse emission of nutrients and water quality in rural areas, mainly in large catchments [12]. Therefore, there is still a paucity of SWAT research on predicting nutrient discharge in small catchments.

In this context, this research provides the results of a study of nitrate losses from the Corbeira catchment, which drains a small (16 km^2) rural landscape in Galicia, NW Spain. The specific objectives of the research were to (i) evaluate the performance and capacity of the eco-hydrological SWAT model to predict nitrate discharge in the Corbeira catchment at monthly time step, (ii) determine the contribution of land use on nitrate losses from this catchment and (iii) assess the relative importance of different pathways by which nitrate is delivered to the drainage network.

The Corbeira catchment was selected because it is representative of the rural landscape in NW Spain, characterized by a distinctly mixed use of the territory, divided between cultivation, pasture and forest. In addition, it is located upstream of the Cecebre reservoir, which is the main water supply for the city of A Coruña and surrounding municipalities (450,000 inhabitants) in northwest Spain. Moreover, this reservoir was declared a Special Area of Conservation and Site of Community Importance included in the Natura 2000 Network. Also, a large number of measurements and analyses related to diffuse pollution have been conducted in this catchment since 2004, which reduces model uncertainty.

2. Material and methods

2.1. Study area and data

The study area was the Corbeira catchment, a small catchment (16 km^2) located at about 30 km northwest of the city of A Coruña (Galicia, NW Spain; **Figure 1**), at a latitude of 43° 13′2.3″N and a longitude of 8° 13′43.9″W. The elevation in the catchment ranges from 47 m at the outlet of the catchment to 470 m at the highest peaks. The topography is moderately steep with an average slope of 19%, and on some areas of the catchment, the slope gradient can reach more than 55%. The geology of the drainage is homogeneous and is dominated by basic schists of the Órdenes Complex [13]. The main types of soil are Umbrisols and Cambisols [14] with great depth and silt and silt-loam texture. The distribution of land cover in the study area is as follows: 65% forest (mainly commercial eucalyptus plantations), 26% pastures and 4% croplands (maize and winter cereal), with the remaining land occupied by impervious areas

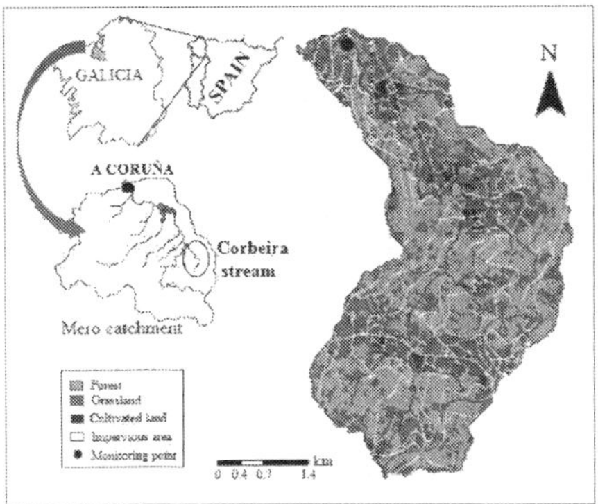

Figure 1. Location of the study area and land use distribution in the Corbeira catchment.

(built-up and infrastructure) mainly distributed in the agricultural zone. Organic and inorganic fertilizers are commonly applied to the agricultural area throughout the year, including the wettest months. However, forest areas are not fertilized. The annual N input to the catchment is approximately 37.8 kg N ha^{-1}, 49% comes from organic fertilizers, 16% from inorganic fertilizers, 2% from population centres and the remaining 33% of atmospheric deposition [15, 16].

The prevailing climate in the study area is humid temperate. The mean annual rainfall is about 1050 mm (historic series: 1983–2009), distributed evenly throughout the year. The mean temperature is 13°C, and the mean annual discharge is around 0.20 m^3 s^{-1} [17].

2.2. SWAT model description

The eco-hydrological SWAT model is a process-based, spatially semi-distributed and continuous model that operates at daily intervals [10]. It was developed by the Agricultural Research Service of the United States Department of Agriculture (USDA) to quantify and predict the impact of agricultural management practices on water, sediment and chemical yields in large complex catchments [10, 11].

In the model, the watershed is divided into sub-basins connected by a stream network. Each sub-basin, in turn, is separated into Hydrological Response Units (HRUs), i.e., territorial units characterized by a specific combination of land use, soil type and slope. The model considers each HRU to be homogeneous in terms of vegetation growth, processes of generation of runoff, erosion and nutrient loading, so they are useful to discriminate the main water, sediment and nutrient sources within each sub-basin. SWAT simulates each HRU separately and the results from HRU_S are integrated at sub-basin scale. It is assumed that there is no interaction between HRUs. The model is flexible in the discretization of the watershed, allowing the user to choose the outlet of the sub-basin. This makes it possible to obtain results of water quantity and quality for any previously selected point, which usually coincides with monitoring stations.

SWAT model simulations are divided into two parts. The first part (land phase) is related to the amount of water, sediment and nutrients delivery from each HRU to the main channel in each sub-basin. The second part (water phase) is related to the behaviour of water and other elements through the channel to the catchment outlet. The model simulates the nitrogen cycle in soil and groundwater, taking into account denitrification, nitrification, mineralization, volatization and plan uptake. SWAT distinguishes five different pools of nitrogen in the soil: two pools are inorganic forms of N and the other three are organic forms of nitrogen. Nitrate is transported from upland areas to stream network via surface runoff, lateral flow and groundwater flow. The amount of nitrate transported by water is calculated by multiplying the nitrate concentration in the mobile water by the volume of water moving in each pathway. Additional informational about the SWAT model can be found in [11].

2.3. SWAT model set-up and input data

The ArcSWAT version 2009.93.7b was used to create the input files for SWAT. The input data and the sources used to create the SWAT model setup of the Corbeira catchment are summarized in **Table 1**. The catchment outlet was set at the Corbeira catchment gauging station, where the hydrological and water quality data are measured. The DEM was used to delimit the catchment, delineate the stream network in the study area and to obtain the topographic parameters, such as slope gradient and slope length; and stream network characteristics, such as channel slope, length and width. Slope, soil and land use data were used for model parameterization, resulting in 12 HRUs. Slope was divided into three classes (0–13%, 13–25% and >25%) following the FAO classification. Seven soil types were identified according to IUSS Working Group WRB [14] classification. The major land uses were defined as 65% forest, 26% pasture, 4% croplands and 5% impervious. The land use pasture and cropland (maize) were parameterized based on the SWAT land use classes, using the SWAT plant codes *meadow bromegrass* and *corn* to represent pasture and maize land covers, respectively, while a new land use was created for the eucalyptus forest area, based on literature [18, 19]. Several management operations (e.g., planting, harvesting and fertilization) were applied for maize and pastures based on knowledge of crop management practice in the catchment and on interviews with the farmers in the catchment.

Data type	Data description	Source
Topography	Digital Elevation Model (DEM), resolution (7 m × 7 m)	Xunta of Galicia
Soils	Soil types (1:50,000)	Xunta of Galicia
Land use	Land use classification	Landsat satellite images provided by Xunta de Galicia
Climate	Daily rainfall, maximum and minimum temperatures, relative humidity, solar radiation and wind speed	Galicia Meteorological Service

Table 1. Model input data sources for the Corbeira catchment.

2.4. Data used for calibration and validation

SWAT was calibrated and validated against nitrate load data measured at the catchment outlet. Nitrate load was calculated as the sum of the product of the mean concentration of adjacent samples by the cumulative flow for each interval of time between both samples. Stream discharge was calculated based on water levels recorded at a 10-min frequency and the level-discharge rating curve and was summarized into mean daily discharge. Water samples for nitrate determination were collected manually with sampling intervals of 10 and 15 days and more intensively (2–8 h) during runoff events, for which an automatic sampler (ISCO 6712) was used, with storage capacity for 24 one-litre polyethylene bottles. The sampler was programmed to begin sampling with increases of 2–3 cm above the stream water level at the beginning of each rainfall event. Nitrate concentrations were analysed by capillary electrophoresis. Model calibration was performed manually on a monthly time step to obtain a reasonably good agreement between the observed and simulated values. The simulation period was limited to five hydrological years (2005/2006–2009/2010) because of data availability; the first three years (2005/2006–2007/2008) were used for calibration and the last two

(2008/2009–2009/2010) for validation. Prior to calibration, the model was warmed up (March 2001–October 2005) to minimize the effect of uncertain initial conditions.

2.5. Model performance evaluation

The model performance was evaluated using the following statistical indices: determination coefficient (R^2), the percent bias (PBIAS) and the Nash-Sutcliffe efficiency (NSE). Information on the statistical equations and the goodness fit of a model at a different time step can be found in [20]. The recommended values for attaining a good model performance for nutrient yield simulations at a monthly time step are PBIAS between ±15 and ±30% and NSE between 0.50 and 0.60.

3. Results and discussion

3.1. SWAT model performance for nitrate yield estimation at catchment scale

The SWAT model used in this research was satisfactorily applied for simulating stream discharge in the catchment under study [21]. In this study, only the parameters that significantly affected nitrate loads such as nitrogen percolation coefficient (NPERCO), humus mineralization (CMN) and residue mineralization (RSDCO) were manually adjusted to provide a good fit between measured and simulated NO_3-N loads. **Table 2** shows the model results based on the performance indicators included in the study, whereas **Figure 2** illustrates the comparison between the measured and simulated monthly nitrate load at the catchment outlet. The simulated mean monthly values were close to the observed values during both the calibration and validation periods, with PBIAS within the 6% of measured NO_3-N load. In general, model simulations can be considered satisfactory ($R^2 > 0.5$; NSE > 0.5) according to the criteria given by Moriasi et al. [20], indicating that it is a valid tool to identify crucial pollution areas within the catchment.

	R^2	PBIAS	NSE
Calibration period	0.52	3	0.50
Validation period	0.54	6	0.53

R^2: regression coefficient, PBIAS: percentage of bias; NSE: Nash-Sutcliffe efficiency

Table 2. Calibration and validation statistics for monthly nitrate yield.

Although simulated NO_3-N yield replicated the measured data trend quite well (**Figure 2**), the model underestimated the measured values during the autumn-winter 2005/2006 when high nitrate levels in stream were observed [16]. This fact may be due to underestimation of some discharge peaks in this period [21], which led to underestimating corresponding NO_3-N yield because NO_3-N, like other water quality parameters, depends on hydrological processes, and therefore errors in discharge simulations are magnified in their simulation. Other authors (e.g.,

[3, 4, 23]) using the SWAT model also attributed the unsatisfactory NO_3-N simulations to problems in discharge simulations. This should be considered a weakness of SWAT to perform NO_3-N simulations at high discharge rates, especially in small streams, such as Corbeira, due to wide variations of discharge and nutrient concentrations during runoff events.

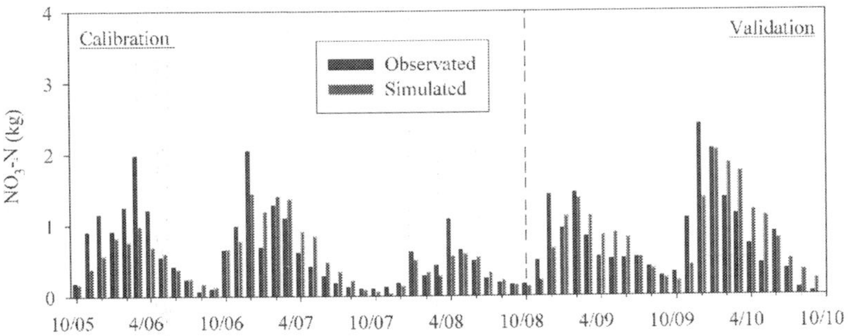

Figure 2. Measured and estimated monthly nitrate yield during the study period. Black broken line marks the separation between the calibration and validation periods.

Annual NO_3-N yield showed high inter-annual variability, ranging from 5.6 kg ha^{-1} in 2005/2006 to 2.8 kg ha^{-1} in 2007/2008. The mean annual NO_3-N measured at the Corbeira catchment outlet was 4.8 kg ha^{-1}, which is comparable with the mean simulated value of 5.1 kg ha^{-1}. The model simulates the NO_3-N yield at monthly and annual scale with sufficient accuracy, indicating that the SWAT model is an appropriate tool for simulating nitrate discharge under the conditions prevailing in the Corbeira catchment. It could be a useful tool for predicting the effect of land use and climate change on NO_3-N yield. Its usefulness could be extended to evaluating management plans aimed at implementing the Water Framework Directive [8] in the Corbeira catchment and in areas with similar environmental and geomorphological conditions. For such purposes, it is essential nutrient yield to be estimated precisely, otherwise results from the model become more uncertain [9], with the consequent impact on environmental management plans. In the Corbeira catchment, it was found that NO_3-N yield, and especially those of particulate phosphorus, exhibited wide variability depending on the sampling technique, method of calculation and evaluation period. It was also seen that the monthly and fortnightly sampling widely underestimated nutrient loads, mainly particulate phosphorus [24, 25]. If data loads not reflecting reality are used, model calibration will be guided to the incorrect setting, as was shown by Ullrich and Volk [9] and Rodríguez-Blanco et al. [25], among others.

3.2. Nitrate yield from the different land uses

Nitrate yield is strongly influenced by land uses within a catchment. To evaluate the diffuse sources of pollution and quantify the nitrate yield entering the catchment from different land uses, the contribution of these to nitrate yield was investigated. This will help to identify the critical land use type and areas for nitrate loss, which is of vital importance in designing catchment management plans aimed at reducing nitrate losses in this type of landscape. **Figure 3** shows the spatial distribution of nitrate yield at HRU scale in the Corbeira catchment. It was observed that the spatial patterns of NO_3-N yield varied significantly in the catchment. The highest nitrate yields were recorded in cultivated areas (17.6 kg ha^{-1} $year^{-1}$) and pastures (13.5 kg ha^{-1} $year^{-1}$), whereas the lowest values were obtained in forest areas (1.98 kg ha^{-1} $year^{-1}$). These results clearly indicate that NO_3-N loss increases with agricultural land use. In fact, cultivated land exported 1.2 and 9 times more nitrate than pastures and forest lands, respectively. The simulated results were comparable to the values reported by other authors in agricultural catchments. Thus, Ferrant et al. [3] found NO_3-N losses of 13 kg ha^{-1} $year^{-1}$ in a small, intensive agricultural catchment in France. Similarly, Frink [26] indicated NO_3-N losses of 3.5–15 kg ha^{-1} $year^{-1}$ from pastures.

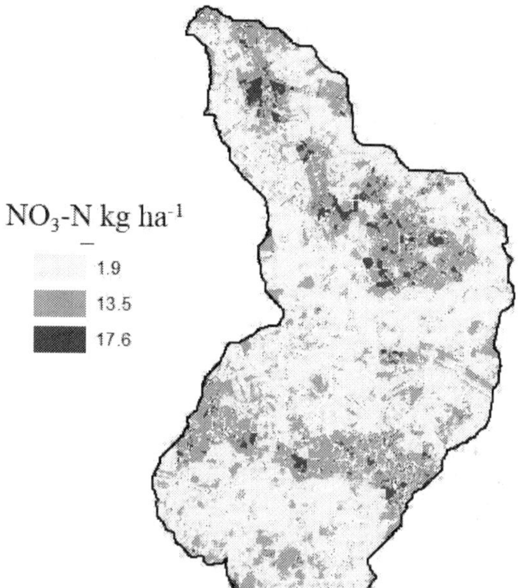

Figure 3. Spatial distribution of nitrate yield in the Corbeira catchment.

The analysis of the contribution of land uses to NO_3-N yield revealed that agricultural lands (pastures and cultivated lands), despite representing only 30% of the catchment area, were the dominant contributor to the total nitrate yield (77%), whereas the forest area (65% of the catchment area) had little influence on nitrate yield in comparison with the other land use. This is mainly due to agricultural land receiving a supply of nitrogen fertilizers (mostly slurry) significantly higher than that of forest land. Therefore, measures aimed at reducing nitrate losses in the catchment should focus on agricultural areas, especially in the pastures, since they are the main area source of NO_3-N area in the Corbeira catchment (64% of total nitrate yield). The above results show the usefulness of SWAT to evaluate the spatial distribution of nitrate yield and identify the most sensitive areas to nitrate pollution within the catchment. Therefore, it would be a very useful tool for evaluating the influence of alternative management practices in controlling nitrate losses, which is in line with the requirements of WFD.

3.3. Transport pathways of nitrate yield

To evaluate the major processes controlling nitrate transport, the contribution of flow component to nitrate yield was evaluated. It was observed that in the Corbeira catchment about 63% of the nitrate load was transported in groundwater, 27% in lateral flow and the remaining 10% via surface runoff. These results are in accordance with previous studies by Rodríguez-Blanco et al. [16] who found that groundwater is the dominant pathway for nitrate in the study catchment, accounting for 60% of total nitrate losses. These results are also in agreement with the studies by Lam et al. [4] and Hu et al. [27], among others, who reported that the groundwater was the major pathway for NO_3-N in areas located in humid zones. These results suggest that management practices aimed at reducing the nitrate load from agriculture in the Corbeira catchment, and in other areas with similar climate, geomorphological and land use characteristics, should mainly focus on reducing the NO_3-N leaching in the catchment.

4. Summary and conclusions

The results from the present study showed an existence of an agreement between measured and model estimations of nitrate yield at catchment outlet, although SWAT underestimated the measured values during some months in the calibration period. The mean annual measured NO_3-N yield was 4.8 kg ha^{-1}, whereas the mean annual simulated NO_3-N yield was 5.1 kg ha^{-1}. The model performance was satisfactory ($R^2 > 0.5$; NSE > 0.5 and PBIAS < 10%) in both calibration and validation periods, indicating that SWAT is an appropriate tool for simulating nitrate discharge under the conditions prevailing in the Corbeira catchment and, consequently, in catchments of similar topography, soil, land use, climate and management. A large spatial variability in the NO_3-N yield was observed within the catchment. As expected, cultivated land had the highest NO_3-N loss (17.6 kg ha^{-1}) and the forest had the lowest (1.98 kg ha^{-1}), indicating that NO_3-N loss increases with agricultural land use. Agricultural land (pasture + cultivated land) accounted for 77% of the NO_3-N losses although they represent only 30% of the catchment area, the pasture being the major contributor of total NO_3-N yield (64%). The results also indicated that groundwater was the major nitrate transport pathway within the catchment, accounting for 63% of the total NO_3-N yield. Lateral flow and surface runoff accounted for 27% and 10% of the NO_3-N yield, respectively. Based on these results, management practices in this catchment should be focused on reducing the leaching of nitrate from agricultural land.

Acknowledgements

This chapter is a contribution to the projects 10MDS103031 of the Xunta of Galicia and CGL2014-56907-R of the Programa Estatal de Investigación, Desarrollo e Innovación Orientada a los Retos de la Sociedad funded by the Spanish Ministry of Economy and Competitiveness. The first author was a recipient of a post-doctoral research contract (Ángeles Alvariño) funded by the Xunta of Galicia and now is beneficiary of a "Juan de la Cierva" research contract supported by the Spanish Ministry of Economy and Competitiveness.

Author details

Maria-Luz Rodríguez-Blanco[1*], Ricardo Arias[1], Maria-Mercedes Taboada-Castro[1], Joao Pedro Nunes[2], Jan Jacob Keizer[2] and Maria-Teresa Taboada-Castro[1]

*Address all correspondence to: mrodriguezbl@udc.es

1 Faculty of Sciences, Centre for Advanced Scientific Research (CICA), University of A Coruna, A Coruña, Spain

2 CESAM and Department of Environment and Planning, University of Aveiro, Aveiro, Portugal

References

[1] Forman RTT, Godron M. Landscape ecology. New York: John Wiley & Sons; 1986. 620 pp.

[2] Macías F, Otero JL, Romero E, Verde R, Parga E, Rodríguez L, Macías F, Taboada M. Monitoring of soil and water pollution of Galicia by eutrophication agricultural residues. Santiago de Compostela: Xunta de Galicia; 2003. 207 pp

[3] Ferrant S, Oehler F, Durand P, Ruiz L, Salmon-Monviola J, Justes E, Dugast P, Probst A, Probst JL, Sanchez-Perez JM. Understanding nitrogen transfer dynamics in a small agricultural catchment: Comparison of a distributed (TNT2) and a semi distributed (SWAT) modeling approaches. Journal of Hydrology. 2011;40:1–15. DOI: 10.1016/j.jhydrol.2011.05.026.

[4] Lam QD, Schmalz B, Fohrer N. Assessing the spatial and temporal variations of water quality in lowland areas, Northern Germany. Journal of Hydrology. 2012;428–439:137–147. DOI: 10.1016/j.jhydrol.2012.03.011.

[5] Howden NJK, Bowes MJ, Clarka ADJ, Humphriesa N, Neal C. Water quality, nutrients and the European union's Water Framework directive in a lowland agricultural region:

Suffolk, south-east England. Science of the Total Environment. 2009;407:2966–2979. DOI: 10.1016/j.scitotenv.2008.12.040.

[6] Kronvang B, Bechmann M, Lundekvam H, Behrendt H, Rubaek GH, Schoumans OF, Syversen N, Andersen HE, Hoffmann CC. Phosphorus losses from agricultural areas in river basins. Journal of Environmental Quality. 2005;34:2119–2144.

[7] EEA (European Environment Agency). Analysis and mapping of soil problem areas (hot spots). Final report. Denmark: Environmental European Agency; 2001. 62 pp.

[8] EC (European Commission). European Parliament and the Council of the European Union. Directive 2000/60/EC establishing a framework for the Community action in the field of water policy. Official Journal of the European Communities. L327; 2000:1–72.

[9] Ullrich A, Volk M. Influence of different nitrate-N monitoring on load estimation as a base for model calibration and evaluation. Environment Monitoring Assessment. 2010;171:513–527. DOI: 10.1007/s10661-009-1296-8.

[10] Arnold JG, Srinivasan R, Muttiah RS, Williams JR. Large area hydrologic modeling and assessment. Part I: Model development. Journal of American Water Resources Association. 1998;34:73–89.

[11] Neitsch SL, Arnold JG, Srinivasan R, Williams JR. Soil and water assessment tool theoretical documentation version 2009. Texa: Grassland, Soil and Water Research Laboratory, Blackland Research Center; 2011. 647 pp.

[12] SWAT Database. Literature Database for Peer-Reviewed Journal Articles [Internet]. 2014. Available from: https://www.card.iastate.edu/swat_articles/.

[13] IGME (Instituto Tecnolýgico Geominero de España). Mapa Geolýgico de España, 1:50,000. Hoja 45. Betanzos; 1981.

[14] IUSS Working Group WRB. World Reference Base for Soil Resources 2014, update 2015. International soil classification system for naming soils and creating legends for soil maps. World Soil Resources Reports. No. 106. Rome: FAO; 2015.

[15] Diéguez A, Rodríguez-Blanco ML, Taboada-Castro MM, Taboada-Castro MT. Interrelationships of water quality, nutrients and major ions in a stream draining a mixed-use catchment, NW Spain. Communications in Soil and Plant Analysis. 2012;43:95–101. DOI: 10.1080/00103624.2011.638532.

[16] Rodríguez-Blanco ML, Taboada-Castro MM, Taboada-Castro MT, Oropeza-Mota JL. Relating nitrogen export patterns from a mixed land use catchment in NW Spain with rainfall and streamflow. Hydrological Processes. 2015;29:2720–2730. DOI: 10.1002/hyp. 10388.

[17] Rodríguez-Blanco ML, Taboada-Castro MM, Taboada-Castro MT. Rainfall runoff response and event-based runoff coefficients in a humid area (northwest Spain). Hydrological Science Journal. 2012;57(3):445–459. DOI: 10.1080/02626667.2012.666351.

[18] Nunes JP, Seixas J, Pacheco NR. Vulnerability of water resources, vegetation productivity and soil erosion to climate change in Mediterranean watersheds. Hydrological Processes. 2008;22:3115–3134. DOI: 10.1002/hyp.6897.

[19] Rodríguez-Suárez JA, Soto B, Iglesias ML, Díaz-Fierros F. Application of the 3PG forest growth model to a Eucalyptus globulus plantation in Northwest Spain. European Journal Forest Research. 2010;129:573–583. DOI: 10.1007/s10342-010-0355-6.

[20] Moriasi DN, Arnold JG, van Liew MW, Bingner RL, Harmel RD, Veith TL. Model evaluation guidelines for systematic quantification of accuracy in watershed simulations. Translation ASABE. 2007;50:885–900.

[21] Rodríguez-Blanco ML, Taboada-Castro MM, Taboada-Castro MT. Simulation of stream discharge from an agroforestry catchment in NW Spain using the SWAT model. In: Proceedings of 2012 International SWAT Conference, 16–20 July 2012; New Delhi, India; 2013. pp. 672–682.

[22] Ferrant S, Oehler f, Durand P, Ruiz L, Salmon M, Justes E, Dugast P, Probst A, Probst JL, Sanchez-Perez JM. Understanding nitrogen transfer dynamics in a small agricultural catchment: comparison of a distributed (TNT2) and a semi distributed (SWAT) modeling approaches. Journal of Hydrology. 2011;406:1–15. DOI: 10.1016/j.jhydrol.2011.05.026.

[23] Yevenes MA, Mannaerts CM. Seasonal and land use impacts on the nitrate budget and export of a mesoscale catchment in Southern Portugal. Agriculture Water Management. 2011;102:54–65. DOI: 10.1016/j.agwat.2011.10.006.

[24] Rodríguez-Blanco ML, Taboada-Castro MM, Taboada-Castro MT. Phosphorus transport into a stream draining from a mixed land use catchment in Galicia (NW Spain): significance of runoff events. Journal of Hydrology. 2013;481:12–21. DOI: 10.1016/j.jhydrol.2012.11.046.

[25] Rodríguez-Blanco ML, Taboada-Castro MM, Palleiro L, Taboada-Castro MT. Analysing the influence of different monitoring strategies on nutrient yield estimation from a small rural catchment in NW Spain. In: Transboundary Water Management Across Borders and Interfaces: Present and Future Challenges, 16–20 March 2013, Aveiro; 2013. pp. 1–4.

[26] Frink CR. Estimating nutrient exports to estuaries. Journal of Environmental. 1991;20:717–724.

[27] Hu X, McIsaac GF, David MB, Louwers CAL. Modeling riverine nitrate export from an east-central Illinois watershed using SWAT. Journal of Environmental Quality. 2007;36:996–1005. DOI: 10.2134/jeq2006.0228.

Section 2

Landscape District in Applied Ecological Analysis

Chapter 3

Multitemporal Analysis in Mediterranean Forestland with Remote Sensing

Ignacio Melendez-Pastor, Encarni I. Hernández,
Jose Navarro-Pedreño, Ignacio Gómez and
Magaly Koch

Additional information is available at the end of the chapter

http://dx.doi.org/10.5772/63721

Abstract

The study employs a Fourier transform analysis approach to assess the land-cover changes in a mountainous Mediterranean protected area using multi-temporal satellite images. Harmonic analysis was applied to a time series of Landsat satellite images acquired from 1984 to 2008 to extract information about land cover status with a vegetation spectral index, the Normalized Difference Vegetation Index (NDVI). Ancillary cartographic information depicting land cover classes and the enlargement of the protected area over time (i.e., maps showing the original delineation in 1995 and subsequent enlargement in 2007) were employed as additional factors to understand vegetation-cover changes. Significant differences in the NDVI and harmonic components values were observed with respect to both factors. The application of the Fourier transform was particularly successful to extract subtle information. The harmonic analysis of the NDVI time series revealed valuable information about the evolution of the landscape. The initially protected area (northern sector) seems more affected by human activities than the southern sector (enlarged area in 2007) as revealed by the analysis of the first harmonic component that was closely related with vegetation coverage. Rural abandonment is a major driver of land-cover changes in the study area.

Keywords: landscape dynamics, forest management, rural abandonment, remote sensing, Fourier transform

INTECH
open science | open minds

1. Introduction

Land cover (i.e., biophysical attributes of the Earth's surface) changes play an important role in the global environmental changes [1,2]. Land-cover changes may result from human-induced land-use changes or natural processes such as climatic variability and natural disturbances [3]. Studying land-cover change is a challenge because it represents a global dynamic phenomenon affecting the Earth's large surface areas and continuously evolves throughout time. Fortunately, the status and rate of land-cover changes may be estimated and monitored through the analysis of remotely sensed data [4]. Remote sensing allows the study of the role of terrestrial vegetation in large-scale global processes (e.g., the carbon cycle) [5]. The scientific community often relies on data gathered by satellite sensors because of their synoptic view, worldwide coverage, and repeated temporal sampling capability of the Earth's surface enabling monitoring of vegetation dynamics from regional to global scales [6,7].

Optical multispectral remote sensing sensors record the reflected portion of electromagnetic radiation-illuminating land covers. They record reflected light as digital counts that can be later transformed into comparable physical measurements such as surface reflectance, which facilitates the comparison with spectrometer measurements of surface covers. Many data transformation techniques (i.e., mathematical operations to reexpress the information content of multispectral and hyperspectral data) exist to perform spectral comparison across sensor platforms, which greatly facilitate the interpretation of land cover types [8]. In this context, vegetation indices have been used to derive measures that correlate with surface biophysical properties [9] and as such facilitate the analysis of large amounts of satellite data for spatial- and temporal-scale analyses [10]. Vegetation indices are directly related to plant vigor, density, and growth conditions and can be used to detect environmental conditions and human activities [4,11]. The most frequently used vegetation index is the Normalized Difference Vegetation Index (NDVI) as described by Rousse et al. [12], which is employed in this study.

Vegetation indices are frequently employed for the characterization of vegetation dynamics and landscape changes using satellite time-series data. They are fundamental data for monitoring vegetation phenology or land-cover changes associated with events such as fire, drought, land-use conversion, and climate fluctuation [13,14]. However, using vegetation-index time series for the analyses of vegetation changes involves a highly complex task and requires the application of specific techniques and methodologies that have been developed over time [11]. The application of Fourier Transform (FT) technique (harmonic analysis) facilitates the extraction of valuable and interpretable characteristics from the time series, which are usually distorted by atmospheric noise, sensor instability, and/or orbit deviations [15]. Harmonic analysis of vegetation-index time series can be used to analyze changes in land covers and vegetation status by examining the altered values of amplitude, phase, or the additive term over a period of years [11,16].

The objective of this study was the evaluation of landscape dynamics by the Fourier Transform technique for the analysis of land-cover changes using a satellite image time series. Harmonic analysis was applied to a time series of Landsat satellite images acquired from 1984 to 2008 to extract information about landscape dynamics using a vegetation index. Landscape dynamics were assessed by combining harmonic analysis results with ancillary cartographic information about the types of land cover classes of the protected area.

2. Material and methods

The study area is located in the *Rodeno* pine forest protected reserve (*Paisaje Protegido de los Pinares de Rodeno* or PPPR in Spanish) of the southern Iberian System in the Sierra de Albarracín shire (Teruel province, Spain). The study area is positioned around 40.33 °N and 1.36°W. The PPPR was established in 1995 covering an area of 3355.34 ha (**Figure 1**). In 2007, the protected landscape was enlarged to about 6829.05 ha. The PPPR occupies part of the municipalities of Albarracín, Gea de Albarracín and Bezas, with a current population (2015) of 1049, 387, and 67 inhabitants, respectively.

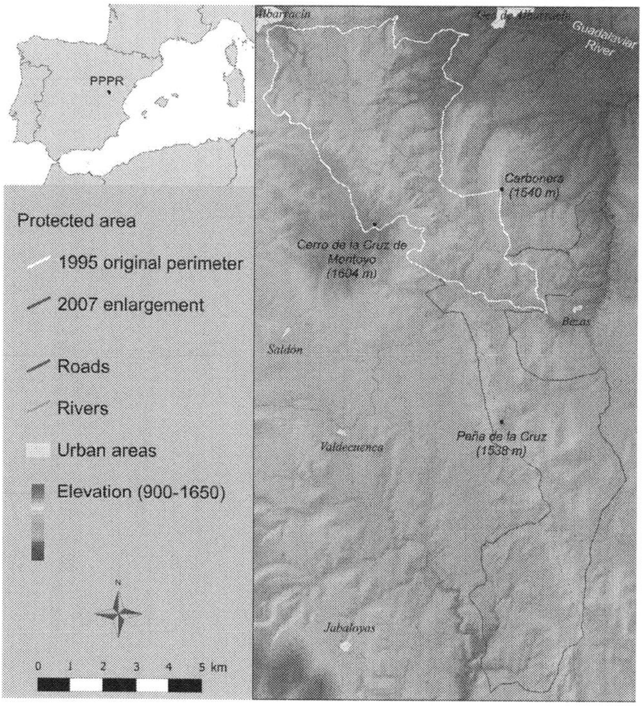

Figure 1. Location map of the study area, the Rodeno pine forest protected reserve (PPPR). Original and enlarged perimeters of the protected area are shown.

It is a mountainous area with maximum altitudes ranging from 1063 to 1601 m and a mean altitude of 1315 m. The climate is characterized by significant seasonal temperature variations between summer (up to 35°C) and winter (down to −22°C). It is classified as wet sub-Mediterranean with an average annual temperature of 9–10°C and annual precipitation of 600–700 mm [17].

The name *Rodeno* is attributed to the characteristic Permian–Triassic reddish sandstones and conglomerates found in this area (**Figure 2a**). Other important lithological materials are Ordovician quartzite and Jurassic limestone and dolomites. Ordovician materials were affected by Hercynian orogeny, and the totality of the study area was shaped by the Alpine orogeny [18,19]. Lithology is an important factor affecting plant species distribution. Extensive *Pinus pinater* Arr. forest (**Figure 2b**) dominates in sandstones and quartzite areas (i.e., more acid soils), while *Quercus ilex* L. and *Juniperus thurifera* L. are more prone to calcaric soils (i.e., soils developed from limestone and dolomite parent materials). In addition to the environmental value of PPPR, the protected area contains an interesting cultural heritage. The presence of numerous rock paintings is particularly noteworthy, which are fortunately protected by law and are recognized as a world heritage by UNESCO (Rock Art of the Mediterranean Basin on the Iberian Peninsula URL: http://whc.unesco.org/en/list/874)

Figure 2. Pictures of the study area: (a) north view from the Peña de la Cruz peak (see location in Figure 1) and (b) characteristic view of a *Pinus pinaster* forest.

Geographical information systems (GIS) and remote-sensing techniques are excellent tools for the identification and analysis of landscape spatial patterns [20]. Thus, satellite imagery and cartographic data were analyzed with digital image processing methods and spatial analysis techniques to detect spatial–temporal land-cover changes. A multitemporal Landsat satellite dataset formed the basis for the change detection procedure. Images were acquired by the multispectral Thematic Mapper (TM) and Enhanced Thematic Mapper Plus (ETM+) sensors onboard Landsat satellites. Images were selected at regular intervals of 8 years from 1984 to 2008. The dates of available scenes were 1984-09-21 (Landsat 4-TM), 1992-04-20 (Landsat 5-TM) the only available scene without cloud cover, 2000-08-08 (Landsat 7-ETM+), and 2008-09-15 (Landsat 7-ETM+ SLC-off). An additional Landsat 7-ETM+ SLC-off acquired on 2008-10-01 was employed for filling gaps in the 2008-09-15 Landsat 7 image. Digital image processing procedure included preprocessing of satellite multispectral images to ensure the temporal comparability (spatial and radiometric coherence and filling the gaps of SLC-off images) between scenes, followed by the calculation of a vegetation index and the application of Fourier Transform analysis to assess temporal changes of land covers.

In addition, several cartographic data sources were employed. For instance, the official protected area boundaries were obtained from the Aragon Territorial Information System (SITAR http://sitar.aragon.es) of the Aragon autonomous community government. The data comprised vector perimeters of the protected area in 1995 and after its enlargement in 2007. Furthermore, a land-cover map was derived from the National Forest Map of Spain [21]. Seven land-cover classes were defined according to dominant vegetation classes included in the vector cartography (**Table 1**). *Pinus pinaster* L stands is the most characteristic vegetation type within the protected area.

Land cover class	Description
Pinus sp.	Pine forest dominated by *Pinus pinaster* L at non-calcareous soils. Pine forest at calcareous soils is denoted by the presence of *Pinus nigra* J.F.Arnold stands.
Juniperus thurifera	Pure stands of small trees of *Juniperus thurifera* L developed at calcareous soils.
Quercus ilex	*Quercus ilex* L. stands preferentially developed at calcareous soils. Usually in the presence of *Juniperus thurifera.*
Juniperus mixture	Formations of medium-sized shrubs dominated by *Juniperus thurifera* and *Juniperus communis* L. They have been intensely grazed and are under natural vegetation processes.
Shrubs	Xerophyte shrubs and pastureland at recursively grazed areas.
Rocky	Non-calcareous rocky areas with the presence of small herbs and shrubs such as *Cistus laurifolius* L. or *Erica scoparia* L.
Agricultural	Non-irrigated cereal crops.

Table 1. Descriptions of land cover classes.

2.1. Preprocessing

Satellite image preprocessing included geometric and atmospheric corrections with the aim to ensure the spatial comparability with other data sources and to obtain at-ground reflectance pixel spectra. Various geo-referenced data types were used for the geometric correction: aerial ortho-photos (1 m pixel resolution) and digital cartography (scale = 1:25000). The 2000 Landsat 7 ETM+ scene was selected because of its very good visual quality. It was geometrically corrected using Ground Control Points (GCP) identified on the ortho-photos and cartographic maps. A quadratic mapping function of polynomial fit and the nearest-neighbor resampling method were used for the correction. The nearest-neighbor resampling method was selected because it ensures that the original (raw) pixel values are retained in the resulting output image, which is an important requirement in any change detection analysis [8]. The maximum allowable root mean square error (RMSE) of the geometric correction was less than half a pixel, a reference value frequently cited as acceptable [5,22,23].

Since May 2003, Landsat 7 images suffer a radiometric problem caused by a malfunctioning scan line corrector (SLC), which compensates for the forward motion of the satellite, and subsequent efforts to recover the SLC were unsuccessful. Fortunately, Landsat 7 is still capable of acquiring useful image data with the SLC turned off, particularly within the central portion of any given scene, and various interpolation and compositing schemes have been developed to expand the coverage of useful data [24]. A local linear histogram-matching method using two SLC-off images (i.e., 15th September 2008 and 1st October 2008 Landsat 7-ETM+ scenes) was applied to fill the gaps [25]. A rescaling function is derived after applying a local linear histogram matching in a moving window over each missing pixel. A rescaling function is then used to convert the radiometric values of a single input scene into equivalent radiometric values of the scene being gap filled, and the transformed data are then used to fill the gaps of that scene [26]. The adaptive local linear histogram adjustment algorithm can yield good results if the scenes have comparable seasonal conditions (the scenes were taken within the shortest revisit time of the satellite over the study area, i.e., 16 days) and do not exhibit radical differences in target radiance (i.e., presence of clouds, snow, sun glint, or changes in solar illumination geometry) [25].

Atmospheric correction involves the estimation of the atmospheric optical characteristics at the time of image acquisition [27]. Radiometric calibration was applied prior to the atmospheric correction. The conversion of raw digital numbers (DN_{raw}) of Landsat level 1 (L1) image products to at-satellite radiance values (L_{sat}) required the application of current rescaling values [28] by applying the following expression [24,28,29]:

$$L_{sat} = \left(\frac{L_{MAX\,\lambda} - L_{MIN\,\lambda}}{255} \right) \cdot DN + L_{MIN\,\lambda} \tag{1}$$

Where L_{sat} is the at-satellite radiance [W/(m^2 sr μm)]; $L_{MIN\lambda}$ is the spectral radiance that is scaled to Q_{calmin} [W/(m^2 sr μm)] (Q_{calmin} is the minimum quantized calibrated pixel value, i.e., DN = 0, corresponding to $L_{MIN\lambda}$); $L_{MAX\lambda}$ is the spectral radiance that is scaled to Q_{calmax} [W/(m^2 sr μm)] (Q_{calmax} is the maximum quantized calibrated pixel value, i.e., DN=255, corresponding to $L_{MAX\lambda}$); and DN are digital numbers of the L1 image product. Surface reflectance values (ρ) were computed using the image-based COST method [30] and applied according to Melendez-Pastor et al. [31]. They computed the path radiance (L_p) values using the equation reported in Song et al. [32] and assuming a 1% surface reflectance for dark objects [30,33,34]. The optical thickness for Rayleigh scattering (τ_r) was estimated according to the equation given in Kaufman [27].

2.2. Vegetation index

The spectral reflectance of the vegetated surface covers recorded by remote sensors can be compressed into vegetation indexes [9], which are directly related to plant vigor, density, and growth conditions and may also be used to detect unfavorable environmental conditions [4]. Vegetation indexes provide an effective way to perform valuable large spatial-and temporal-scale analyses with large amounts of satellite data [10]. The normalized difference vegetation index (NDVI) [12] has been frequently employed for vegetation status assessment and is formulated as follows:

$$NDVI = \frac{\rho_{NIR} - \rho_{red}}{\rho_{NIR} + \rho_{red}} \tag{2}$$

Where ρ_{NIR} is the reflectance of the near infrared spectral band and ρ_{RED} is the reflectance of the red spectral band. NDVI has been related with several vegetation parameters such as changes in the amount of green biomass and chlorophyll content, leaf water content, CO_2 net flux, absorbed photosynthetically active radiation (APAR), leaf area index (LAI), and many others [35]. It is also employed as vegetation status data source in many environmental modeling approaches [36,37].

2.3. Fourier transform

Satellite NDVI time series can be analyzed by means of the Fourier Transform in order to obtain its frequency domain components [11]. Harmonic or Fourier analysis allows the decomposition of temporal data to the frequency domain, by which frequency information is represented in terms of the sum of a set of sine and cosine functions [38]. Harmonic analysis allows a complex curve to be expressed as the sum of a cosine waves (terms) and an additive term [39]. The Fourier Transform converts a function $f(x)$ to the frequency domain by a linear combination of trigonometric functions [40].

The correct interpretation of harmonic analysis results requires the consideration of several key concepts [11,41]: (1) each wave is defined by a unique amplitude and phase angle, where amplitude is the difference between the maximum value of a wave and the overall average of the time series (usually 0 because the mean is equal to the harmonic term with $k = 0$), and the phase angle defines the offset between the origin and the peak of the wave over the range 0 to 2π; (2) the additive or zero term is the arithmetic mean of the variable over the time series; (3) high amplitude values for a given term indicate a high level of variation in the temporal variable time series, and the term in which that variation occurs indicates the periodicity of the event; and 4) phase indicates the time at which the peak value for a term occurs. Each term indicates the number of complete cycles completed by a wave over a defined interval [16]. In this study, the Fourier Transform was applied to the 24-year Landsat time series. The additive term and the first two harmonic components of the NDVI time series were extracted for all land-cover classes for subsequent analysis. IDRISI Selva © software (Clark Labs, Clark University) was employed for the harmonic analysis of the time series.

2.4. Statistical analyses

Statistical analyses were used to assess the influence of the time of landscape protection (1995 original perimeter of the protected area vs. 2007 park enlargement) and the land-cover classes on the vegetation cover status as denoted by the NDVI images and derived harmonic components. Data for statistical analysis were obtained by randomly selecting pixels on the NDVI and harmonic components images. A total of 2742 pixels within the current perimeter of the PPPR were obtained to extract the NDVI and harmonic component values, and further statistical analyses were performed with the IBM-SPSS statistics 23 © software (IBM Corporation).

Input variables were checked for normal distribution with the Kolmogorov–Smirnov test. Required variable transformations were applied to ensure the normality of input variables. A one-way analysis of variance (ANOVA) test was employed to detect the NDVI or harmonic component differences using the protected area declaration (PAD) as factor. The same procedure was applied to detect NDVI or harmonic component differences using land-cover classes (LC) as a factor. In addition, a two-way ANOVA test was employed to assess the harmonic component differences using protected area declaration (PAD) and land-cover classes (LC) as factors.

3. Results and discussion

The PPPR study area is a mountainous area with a landscape dominated by the presence of large coniferous forest (*Pinus sp.*). **Figure 3a** shows a false color composite RGB = 742 of the 1984 Landsat TM image. Green areas are associated with coniferous forest, whitish areas correspond to almost bare soils of agricultural fields, and reddish areas correspond to sandstone areas with different vegetation classes. A digital elevation model (DEM) shown in **Figure 1** allowed the topographic characterization of the area. The mean altitude and average slope of the 1995 protected area and the 2007 enlargement area are very similar. The mean altitude and average slope of the PPPR are 1315 m and 13.7°, respectively.

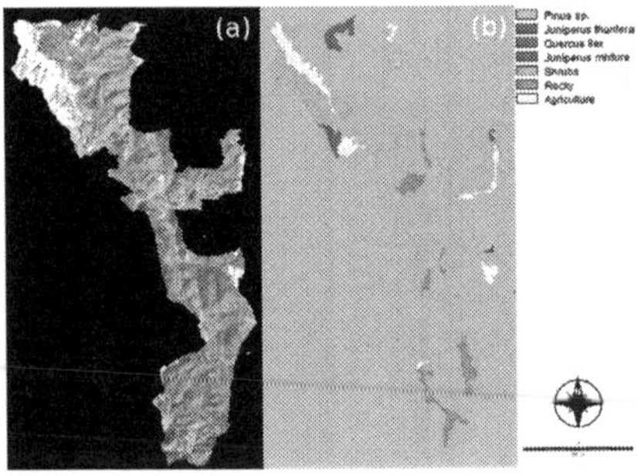

Figure 3. Satellite imagery and cartographic data: (a) false color composite RGB = 742 of the 1984 Landsat satellite image and (b) land cover map from the National Forest Map.

The major land cover class is *Pinus sp.* forest that accounts for 80% (5439.69 ha) of the PPPR (**Figure 3b** and **Table 2**). The proportion was slightly higher for the 2007 enlargement area (88% or 3147.39 ha) compared to that for the 1995 perimeter (70% or 2292.30). Shrub (432.9 ha), rock (405.99 ha), and agriculture (371.52 ha) classes represent about 6% each. *Juniperus thurifera* land cover class is not present in the 1995 perimeter. Meanwhile, *Quercus ilex, Juniperus* mixture, and shrub land-cover classes are no present in the 2007 enlargement.

Land cover clases	Current PPPR (ha)	1995 perimeter (ha)	2007 enlargement (ha)
Pinus sp.	5439.69	2292.30	3147.39
Juniperus thurifera	18.72	No presence	18.72
Quercus ilex	97.02	97.02	No presence
Juniperus mixture	65.43	65.43	No presence
Shrubs	432.90	432.90	No presence
Rocky	405.99	108.72	297.27
Agriculture	371.52	268.65	102.87
TOTAL	**6831.27**	**3265.02**	**3566.25**

Table 2. Land cover surface (ha) distribution in the current PPPR, the original protected area (1995) and 2007 park enlargement.

The average NDVI values for each land-cover class (**Table 3**) were computed for each satellite image (from 1984 to 2008). *Pinus sp.* and rocky areas had the largest average NDVI values with 0.53 and 0.51, respectively. On the contrary, shrubs and agricultural fields had average NDVI values less than 0.36. *Juniperus thurifera, Quercus ilex,* and *Juniperus* mixture stands had sequentially decreasing NDVI values from 0.43 to 0.37.

Variables	$\bar{x} \pm \sigma$	PAD	Land cover
NDVI 1984	0.50 ± 0.31	0.000 ***	0.000 ***
NDVI 1992	0.59 ± 0.35	0.000 ***	0.000 ***
NDVI 2000	0.46 ± 0.28	0.000 ***	0.000 ***
NDVI 2008	0.51 ± 0.32	0.000 ****	0.000 ***
A0	0.51 ± 0.30	0.054 N.S.	0.008 **
A1	0.04 ± 0.59	0.000 ***	0.000 ***
A2	0.05 ± 0.62	0.000 ***	0.000 ***
P1	1.95 ± 1.92	0.060 N.S.	0.220 N.S.
P2	5.30 ± 0.92	0.000 ***	0.000 ***

** = $p \le 0.010$; *** = $p \le 0.001$; N.S. = not significant.
Independent ANOVA test for protected area declaration (PAD) and land cover class factors. Descriptive statistics (mean ± standard deviation) of variables are included. Phase values are in radians.

Table 3. One-way analysis of variance results (*p*-values) for NDVI and harmonic components images.

One-way ANOVA of the NDVI values of the sampled points allowed the independent comparison of the effect of protected area declaration and land-cover classes on vegetation index values (**Table 3**). Significant differences of NDVI values were observed for both factors. Different land cover types tend to report different values of NDVI, thus allowing the identification of temporal patterns of the vegetation index for each land cover type. In addition, the average NDVI values for both protected area declaration (PAD) subzones could be different by their dissimilar landscape structure. These changes in the vegetation index values are related to the phenological changes in the land covers being characteristic of a particular type of vegetation and may be affected by human actions [11]. Several authors have noted the usefulness of vegetation time series for classifying land cover classes in mountainous areas such as the study area [42] and for detecting changes in forest stands [43,44].

The first two harmonic terms and the additive term were derived from the Fourier Transform of the NDVI time series (**Figure 4**). Harmonic analysis allowed the conversion of the NDVI time series from the time domain to the frequency domain, which is very useful to mitigate noise effects and gain information about the time at which NDVI changes occurred. Such information could be obtained after a combined interpretation of the phase and amplitude components. The first three or four harmonic terms concentrate vegetation phenology patterns, while the high-frequency noise is partitioned into the higher harmonic terms [41]. The additive term (A0) corresponds with the average NDVI of the time series.

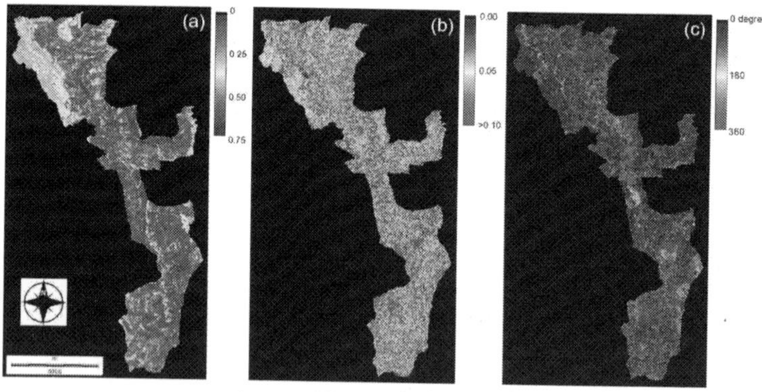

Figure 4. Cartography depiction of NDVI-based Fourier Transforms: (a) additive or amplitude 0 image, (b) amplitude of the first harmonic component, and (c) phase of the first harmonic component.

The average values of the amplitude of the first harmonic component (A1) and the second harmonic component (A2) are shown for each land cover class (**Figure 5a**). Highest amplitude values were observed for *Pinus sp.*, rocky and agricultural land cover classes suggesting larger temporal variations of NDVI values compared to those for the other land cover classes. Conversely, *Quercus ilex* stands seem to be highly stable throughout the time that is particularly important for a low-growth rate species. *Juniperus thurifera* and *Juniperus* mix-

http://dx.doi.org/10.5772/63721

ture land cover classes had moderate A1 values suggesting a progressive but slow regrowth of vegetation due to cessation of rural activities.

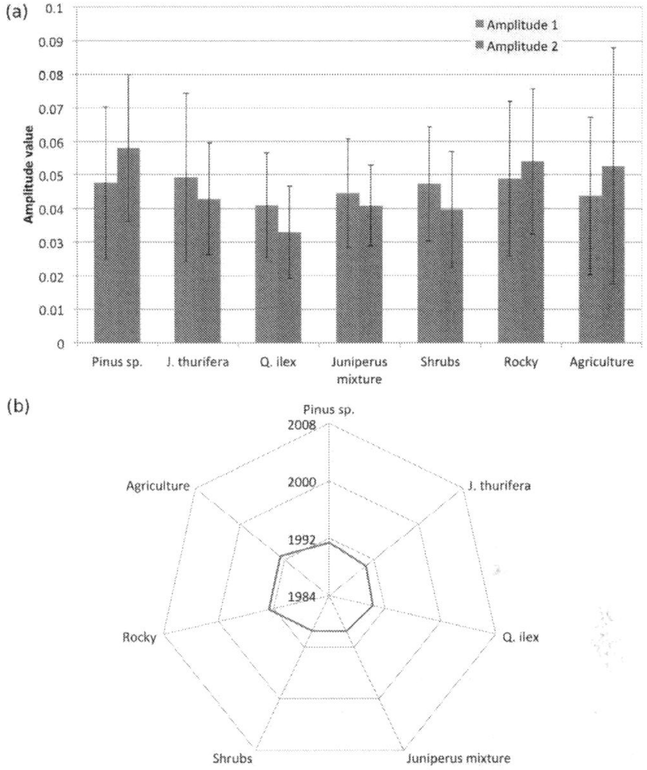

Figure 5. NDVI-based Fourier Transform results: (a) mean value of the amplitude (with standard deviation error bars) for the first two harmonic components for each land cover class; and (b) mean phase value of the first harmonic component phase (in years) for each land cover class.

A two-way ANOVA test with the sampled values of the first two harmonic components and the additive term was performed. Protected area declaration and land cover class were employed as factors. Only land cover classes present in both PPPR protected areas (1995 original area and 2007 enlargement) were considered. A total of 2502 pixel of *Pinus sp.*, rocky and agricultural land cover classes were analyzed (**Table 4**). In relation with the protected area declaration factor, all amplitude and phase variables exhibited significant variations except for the second harmonic component. In relation with the land cover factor, significant differences were observed for zero term and the phase values but not for the other amplitude terms (A1 and A2). Finally, the effects of the interaction of both factors were assessed. Only significant differences were observed for the phase of the first component. A two-way

ANOVA test with NDVI values was discarded because such variables did not show any relevant information. Significant differences of NDVI values for PAD and LC factors were always detected.

Factor	Variable	df	F	*p*-value
PAD	A0	1	14.374	0.000 ***
	A1	1	8.221	0.004 *
	A2	1	2.145	0.143 NS
	P1	1	4.746	0.029 *
	P2	1	1.985	0.159 NS
Land cover	A0	2	260.635	0.000 ***
	A1	2	2.867	0.057 NS
	A2	2	1.661	0.190 NS
	P1	2	14.583	0.000 ***
	P2	2	4.833	0.008 **
PAS I Land cover	A0	2	2.958	0.052 NS
	A1	2	1.297	0.274 NS
	A2	2	1.735	0.177 NS
	P1	2	24.030	0.000 ***
	P2	2	3.697	0.025*

* = $p \leq 0.050$; ** = ≤ 0.010; *** = $p \leq 0.001$; N.S. = not significant.
Protected area declaration (PAD) and land cover class are the factors. Degrees of freedom, F statistic, and *p*-values are included.

Table 4. Two-way analysis of variance results for the harmonic component images.

A final interpretation of the Fourier Transform results was obtained after considering the results of the visual inspection of the images complemented with ancillary geographical information and the two-way ANOVA results. The amplitude of the first harmonic component (A1) was primarily associated with changes in the forest stands by logging and the significant cessation of grazing. We analyzed the A1 image in a GIS and located the highest numeric values areas (dark red areas in **Figure 4b**). Such areas were identified on aerial photography from 1997 and 2008 available at the aforementioned SITAR. We observed a progressive vegetation re-growth of such areas after the cessation of past clear-cutting (currently logging activities are less aggressive, i.e., horses are even used to transport logs to minimize environmental impact) and the progressive decrease of grazing pressure by the rural abandonment. The phase of the first harmonic component (P1; **Figure 4c**) was less than π (in radians or 180 degrees) for such high A1 values suggesting that the peak of the first harmonic component for such areas was before 1996 (**Figure 5**). The average values of the phase 1 component were computed as a function of the land cover classes within the entire PPPR. The peak of the first harmonic component was for the late 1980s and early 1990s with later values for *Pinus sp.*, rocky and agricultural land cover classes. The evolution of the population is a factor commonly used to explain the dynamics of the landscape [45]. The current landscape of the Spanish Mediterranean mountain areas is the result of multiple biophysical agents, such as demographic trends, and the political and economic dynamics [46]. The population of the Sierra de Albarracín is continuously decreasing since the early XX century due to rural abandonment. This process of decline in rural population results in a reduction of an extensive livestock activity that is largely responsible for the maintenance of shrublands and pastures [47] and promoting forest regeneration [7]. In the last 25 years, Spain has undergone deep social and administrative changes (rural abandonment, new forestry policies, and administrative changes) promoting vegetation recovery by the reduction of grazing pressure as a result of depopulation of rural areas [48]. Nevertheless, agricultural and logging activities still persist within the park but are well regulated.

Finally, the application of the Fourier Transform was particularly interesting to extract subtle information. The solely NDVI analysis suggested large temporal and land cover-dependent differences on the vegetation cover, but subtle changes seemed to be undetectable. The harmonic analysis of the NDVI time series revealed highly valuable and easily interpretable information. Higher amplitude values (A0, and A1) were observed for the three land cover classes (*Pinus sp.*, rocky and agricultural) within the 2007 enlargement area. The two-way ANOVA revealed significant differences of the additive term and the first harmonic component (A1 and P1) as a function of the PAD factor (**Table 4**). The additive term also have significant differences as a function of the land cover factor (higher values for the forest cover). The application of the two-way ANOVA was particularly useful to understand the combined effect of land protection and land-cover classes on vegetation dynamics. We suggest that our first harmonic component was a good indicator of vegetation recovery processes. The fact that no significant differences of the A1 component were obtained for the land cover factor but were significant for the PAD factor suggests that the expansion of the park had a positive impact on vegetation coverage in the reclaimed areas. Such differences could not be directly attributable to a poor management of the protected area because the northern sector of the park is closest to the major population centers of the shire (Albarracín and Gea de Albarracín), while the southern sector has greater inaccessibility (worst road network). The presence of a lumber mill in Albarracín (northern study area) suggests larger logging in the northern sector (better accessibility) compared to that in the southern sector of the protected area. The distance to urban areas or roads is a critical factor affecting land-cover changes induced by processes such as deforestation [49,50] land abandonment [51] or urban growth [52].

Future research should be focused on updating the Fourier Transform time series analysis with the inclusion of additional images for September-October 2016 (in order to maintain the 8-year interval). In order to continue the Landsat time-series analysis, Landsat-8 imagery (available from April 2013 on) would be the most appropriate choice as this sensor has similar characteristics to Landsat 7. However, their comparability is somewhat complex by the reflectance differences expected by their notably dissimilar specifications (i.e., narrower bands, 16 bits instead of 8 bits for the OLI (operational land imager) sensor) [53]. It is difficult to comprehensively model their spectral reflectance differences, which also depend on the surface reflectance and atmospheric state [54]. Recent studies have modeled the relationship between the vegetation indices images obtained from ETM+ and OLI [54–56]. This information is very important for future applications of remote sensing to assess landscape temporal dynamics. Vegetated areas had better NDVI agreement than non-vegetated surfaces (especially water areas) [55], and the seasonal agreement of both sensors is better for forest and shrub areas during growth periods than for others land covers [56]. In addition, OLI images are very infrequently saturated [54], and hence more precise quantitative applications of remote sensing are expected for complex landscapes such as wetlands, mountainous areas and forest.

4. Conclusions

This study applied the Fourier Transform technique to analyze land-cover changes based on a NDVI time series of satellite images in a mountain forestland, considering the expansion stages of a protected reserve. Although the complexity of the study area (geomorphology) could affect the expected results, the harmonic analysis provided a highly efficient method to minimize the time-series noise effect and to extract valuable information for interpreting spatial and temporal changes of land covers. The first two harmonic components plus the additive term were selected for the evaluation of the landscape dynamics. Higher amplitude values were obtained for very dynamic land cover classes or drastic landscape changes (e.g., intensive logging areas), while lower amplitude values were associated with slow regeneration vegetation classes. Further statistical analysis revealed significant differences of NDVI harmonic components for the expansion stages of the protected reserve and land-cover factors. Those results suggested a complex landscape dynamics greatly influenced by the land management and vegetation classes. The interpretation of such landscape changes and underlying driving factors was supported by previous research of key socioeconomic and environmental factors. The study area is suffering serious depopulation problems manifested by the progressive expansion of forest areas at the expense of former grazing land. Awareness of environmental and cultural values of the study area has led to the progressive protection of the area to improve conservation and management. In this sense, the use of remote sensing images to explore and understand the temporal evolution of ecosystems was an excellent source of knowledge to improve land management. The Landsat scenes selected to analyze vegetation cover changes between the four time periods in this Spanish Mediterranean mountainous protected area provided valuable information usable to understand landscape dynamics. The combined use of land cover and protected area development maps with the first harmonic components enabled the analysis of landscape dynamics as influenced by rural abandonment. The multitemporal NDVI analysis with the Fourier transform allows assessing the places and the time that logging activities have taken place. The use of these techniques can help improve the study and ecological conservation in remote rural areas throughout the world.

http://dx.doi.org/10.5772/63721

Acknowledgements

The authors would like to thank the U.S. Geological Survey for providing the Landsat satellite imagery used in the study.

Author details

Ignacio Melendez-Pastor[1*], Encarni I. Hernández[1], Jose Navarro-Pedreño[1], Ignacio Gómez[1] and Magaly Koch[2]

*Address all correspondence to: imelendez@umh.es

1 Department of Agrochemistry and Environment, University Miguel Hernández of Elche. Av/ Universidad s/n, Edificio Alcudia. Elche (Alicante), Spain

2 Center for Remote Sensing, Boston University, Boston, Massachusetts, USA

References

[1] IGBP. Land-Use and Land-Cover Change (LUCC) Implementation Strategy. IGBP Report No. 48/IHDP Report No 10. Stockholm, Sweden: International Geosphere-Biosphere Programme; 1999. 126 p.

[2] Sala OE, Chapin III FS, Armesto JJ, Berlow E, Bloomfield J, Dirzo R. Global biodiversity scenarios for the year 2100. Science. 2000;287:1770–1774.

[3] Lupo F, Linderman M, Vanacker V, Bartholomé E, Lambin EF. Categorization of land-cover change processes based on phenological indicators extracted from time series of vegetation index data. International Journal of Remote Sensing. 2007;28:2469–2483.

[4] Li J, Lewis J, Rowland J, Tappan G, Tieszen LL. Evaluation of land performance in Senegal using multi-temporal NDVI and rainfall series. Journal of Arid Environments. 2004;59:463–480.

[5] Jensen JR. Introductory Digital Image Processing (3rd Edition). Upper Saddle River, NJ, USA: Prentice Hall; 2005. 544 p.

[6] Zhang X, Friedl MA, Schaaf CB, Strahler AH, Hodges JCF, Gao F. Monitoring vegetation phenology using MODIS. Remote Sensing of Environment. 2003;84:471–475.

[7] Melendez-Pastor I, Hernández EI, Navarro-Pedreño J, Gómez I. Socioeconomic factors influencing land cover changes in rural areas: the case of the Sierra de Albarracín (Spain). Applied Geography. 2014;52:34–45.

[8] Mather PM, Koch M. Computer Processing of Remotely-Sensed Images: An Introduction (4th Edition). Chichester, UK: Wiley-Blackwell; 2011. 460 p.

[9] Myneni RB, Hall FG, Sellers PJ, Marshak AL. The interpretation of spectral vegetation indexes. IEEE Transactions on Geoscience and Remote Sensing. 1995;33:481–486.

[10] Tucker CJ, Townshend JRG, Goff TE. African land-cover classification using satellite data. Science. 1985;227:369–375.

[11] Melendez-Pastor I, Navarro-Pedreño J, Koch M, Gómez I, Hernández EI. Land-cover phenologies and their relation to climatic variables in an anthropogenically impacted Mediterranean coastal area. Remote Sensing. 2010;2:697–716.

[12] Rouse JW, Hass RH, Schell JA, Deering DW. Monitoring vegetation systems in the great plains with ERTS. In: Proceedings, Third Earth Resources Technology Satellite-1 Symposium. Washington, DC, USA: NASA SP-351; 1974. p. 3010–3017.

[13] Justice CO, Holben BN, Gwynne MD. Monitoring East African vegetation using AVHRR data. International Journal of Remote Sensing. 1986;7:1453–1474.

[14] Moody A, Johnson DM. Land-surface phenologies from AVHRR using the discrete fourier transform. Remote Sensing of Environment. 2001;75:305–323.

[15] Immerzeel WW, Quiroz RA, De Jong SM. Understanding precipitation patterns and land use interaction in Tibet using harmonic analysis of SPOT VGT-S10 NDVI time series. International Journal of Remote Sensing. 2005;26:2281–2296.

[16] Jakubauskas ME, Legates DR, Kastens JH. Crop identification using harmonic analysis of time-series AVHRR NDVI data. Computers and Electronics in Agriculture. 2002;37:127–139.

[17] Government of Aragón. Atlas Climático de Aragón. Zaragoza, Spain: Department of Enviornment Government of Aragón; 2007. 291 p.

[18] IGME. Mapa Geológico de España. E 1:50000. Hoja 566 Cella. Madrid, Spain: Geological and Mining Institute of Spain (IGME); 1983. p. 70.

[19] IGME. Mapa Geológico de España. E 1:50000. Hoja 589 Terriente. Madrid, Spain: Geological and Mining Institute of Spain (IGME); 1983. p. 82.

[20] Gumbricht T, Mccarthy J, Mahlander C. Digital interpretation and management of land cover - a case study of Cyprus. Ecological Engineering. 1996;6:273–279.

[21] MMA. Mapa Forestal de España 1:200.000. Teruel 7-6. Madrid, Spain: Ministry of Environment (MMA); 1991.

[22] Townsend PA, Walsh SJ. Remote sensing of forested wetlands: application of multitemporal and multispectral satellite imagery to determine plant community composition and structure in southeastern USA. Plant Ecology. 2001;157:129–151.

[23] Melendez-Pastor I, Navarro-Pedreño J, Koch M, Gómez I. Applying imaging spectroscopy techniques to map saline soils with ASTER images. Geoderma. 2010;158:55–65

[24] Irish RR. Landsat 7 Science Data Users Handbook. Landsat Project Science Office, Goddard Space Flight Center-NASA [Internet]. 2011. Available from: http://landsathandbook.gsfc.nasa.gov [Accessed: 2016-03-21]

[25] USGS. Phase 2 gap-fill algorithm: SLC-off gap-filled products gap-fill algorithm methodology [Internet]. 2004 Available from: http://landsat.usgs.gov/documents/L7SLCGapFilledMethod.pdf [Accessed: 2016-03-21]

[26] Chen J, Zhu X, Vogelmann JE, Gao F, Jin S. A simple and effective method for filling gaps in Landsat ETM+ SLC-off images. Remote Sensing of Environment. 2011;115:1053–1064.

[27] Kaufman YJ. The atmospheric effect on remote sensing and its corrections. In: Asrar G, editor. Theory and Applications of Optical Remote Sensing. New York, USA: Wiley-Interscience; 1989. p. 336–428.

[28] Chander G, Haque MO, Micijevic E, Barsi JA. A Procedure for radiometric recalibration of Landsat 5 TM reflective-band data. IEEE Transactions on Geoscience and Remote Sensing. 2010;48:556–574.

[29] Chander G, Markham B. Revised Landsat-5 TM radiometric calibration procedures and postcalibration dynamic ranges. IEEE Transactions on Geoscience and Remote Sensing. 2003;41:2674–2677.

[30] Chavez Jr PS. Image-based atmospheric corrections - Revisited and improved. Photogrammetric Engineering & Remote Sensing. 1996;62:1025–1036.

[31] Melendez-Pastor I, Hernández EI, Navarro-Pedreño J, Gómez I. Mapping soil salinization of agricultural coastal areas in southeast spain. In: Escalante B, editor. Remote Sensing—Applications. Rijeka, Croatia: InTech; 2012. p. 117–140.

[32] Song C, Woodcock CE, Seto KC, Lenney MP, Macomber SA. Classification and change detection using Landsat TM data: when and how to correct atmospheric effects? Remote Sensing of Environment. 2001;75:230–244.

[33] Moran MS, Jackson RD, Slater PN, Teillet PM. Evaluation of simplified procedures for retrieval of land surface reflectance factors from satellite sensor output. Remote Sensing of Environment. 1992;41:169–184.

[34] Chavez Jr PS. Radiometric calibration of Landsat Thematic Mapper multispectral images. Photogrammetric Engineering & Remote Sensing. 1989;55:1285–1294.

[35] Jensen JR. Remote Sensing of the Environment: An Earth Resource Perspective. Upper Saddle River, NJ, USA: Prentice Hall; 2007. p. 608

[36] Chen SH, Su HB, Tian J, Zhang RH, Xia J. Estimating soil erosion using MODIS and TM images based on support vector machine and à trous wavelet. International Journal of Applied Earth Observation and Geoinformation. 2011;13:626–635.

[37] Melendez-Pastor I, Navarro-Pedreño J, Koch M, Gómez I, Hernández EI. Evaluation of land degradation after forest fire with a fuzzy logic model. Environmental Engineering and Management Journal. 2013;12:2087–2096.

[38] Briggs WL, Henson VE. The DFT: An Owners' Manual for the Discrete Fourier Transform. Philadelphia, PA, USA: Society for Industrial Mathematics; 1995. 450 p.

[39] Rayner JN. An Introduction to Spectral Analysis. London, UK: Pion Ltd.; 1971. p. 174

[40] Sakamoto T, Yokozawa M, Toritani H, Shibayama M, Ishitsuka N, Ohno H. A crop phenology detection method using time-series MODIS data. Remote Sensing of Environment. 2005;96:366–374.

[41] Jakubauskas ME, Legates DR, Kastens JH. Harmonic analysis of time-series AVHRR NDVI data. Photogrammetric Engineering & Remote Sensing. 2001;67:461–470.

[42] Evans JP, Geerken R. Classifying rangeland vegetation type and coverage using a Fourier component based similarity measure. Remote Sensing of Environment. 2006;105:1–8.

[43] Parent MB, Verbyla D. The browning of Alaska's Boreal Forest. Remote Sensing. 2010;2:2729–2747.

[44] Yen P, Ziegler S, Huettmann F, Onyeahialam AI. Change detection of forest and habitat resources from 1973 to 2001 in Bach Ma National Park, Vietnam, Using Remote Sensing Imagery. International Forestry Review. 2005;7:1–8.

[45] Izquierdo AE, De Angelo CD, Aide TM. Thirty years of human demography and land-use change in the Atlantic Forest of Misiones, Argentina: an evaluation of the forest transition model. Ecology and Society. 2008;13:3.

[46] Gómez Vargas FJ, Boada Juncà M, Sànchez Mateo S. Local models of land use & land cover change analysis. The case of Ridaura sessile oak forestland (Natural Park of Montseny. Barcelona). Boletín de la Asociación de Geógrafos Españoles. 2008;47:379–386.

[47] García-Ruiz JM, Lasanta T, Ruiz-Flano P, Ortigosa L, White S, González C, et al. Land-use changes and sustainable development in mountain areas: a case study in the Spanish Pyrenees. Landscape Ecology. 1996;11:267–277.

[48] Valbuena-Carabaña M, López de Heredia U, Fuentes-Utrilla P, González-Doncel I, Gil L. Historical and recent changes in the Spanish forests: a socio-economic process. Review of Palaeobotany and Palynology. 2010;162:492–506.

[49] Chomitz KM, Gray DA. Roads, land use, and deforestation: a spatial model applied to Belize. The World Bank Economic Review. 1996;10:487–512.

[50] Southworth J, Marsik M, Qiu Y, Perz S, Cumming G, Stevens F, et al. Roads as drivers of change: trajectories across the tri-national frontier in MAP, the Southwestern Amazon. Remote Sensing. 2011;3:1047–1066.

[51] Díaz GI, Nahuelhual L, Echeverría C, Marín S. Drivers of land abandonment in Southern Chile and implications for landscape planning. Landscape and Urban Planning. 2010;99:207–217.

[52] Su S, Xiao R, Zhang Y. Multi-scale analysis of spatially varying relationships between agricultural landscape patterns and urbanization using geographically weighted regression. Applied Geography. 2012;32:360–375.

[53] USGS. Landsat 8 (L8) Data Users Handbook. Version 1.0. Sioux Falls, SD, USA: Earth Resources Observation and Science (EROS) Center. US Geological Survey; 2015. p. 106.

[54] Roy DP, Kovalskyy V, Zhang HK, Vermote EF, Yan L, Kumar SS, et al. Characterization of Landsat-7 to Landsat-8 reflective wavelength and normalized difference vegetation index continuity. Remote Sensing of Environment. 2016;in press. doi: 10.1016/j.rse.2015.12.024

[55] Ke Y, Im J, Lee J, Gong H, Ryu Y. Characteristics of Landsat 8 OLI-derived NDVI by comparison with multiple satellite sensors and in-situ observations. Remote Sensing of Environment. 2015;164:298–313.

[56] She X, Zhang L, Cen Y, Wu T, Huang C, Baig HM. Comparison of the continuity of vegetation indices derived from Landsat 8 OLI and Landsat 7 ETM+ data among different vegetation types. Remote Sensing. 2015:7:13485–13506

Chapter 4

Hydrological Trend Analysis Integrated with Landscape Analysis at the Watershed Scale (Case Study: Langat Basin, Malaysia)

Hadi Memarian and Siva K. Balasundram

Additional information is available at the end of the chapter

http://dx.doi.org/10.5772/62463

Abstract

In this study, the trends of water and sediment data collected from three hydrometer stations over the past 25 years of development in the state of Selangor, Peninsular Malaysia, were analyzed using the Mann–Kendall and Pettitt's tests. Landscape metrics for establishing the relationship between land use changes and trends of hydrological time series were calculated. The hydrologic trends were also studied in terms of rainfall variations and man-made features. Results indicated upward trends in water discharge at the Hulu Langat sub-basin and sediment load at the Semenyih sub-basin. These increasing trends were mainly caused by rapid changes in land use. Upward trends of hydrological series at the Hulu Langat sub-basin matched its rainfall pattern. At the Lui sub-basin, however, trends of hydrological series and variations in rainfall and land use were not statistically significant.

Keywords: trend analysis, Mann–Kendall test, Pettitt's test, landscape metrics, water discharge, sediment load

1. Introduction

Globally, increased sediment load (SL) and intense flooding due to land use changes at river basins are very challenging problems [1–4]. The impacts of human activities and climate change on hydrological processes occurring at river systems are well documented [5–10]. Understanding time series trends of water discharge (WD) and SL can be a key solution in determining how hydrological systems are affected by climate change and anthropogenic disturbances [4].

Zhang et al. [4] determined time series changes in water and sediment discharge at the Zhujiang (Pearl River) Basin in China. They applied Mann–Kendall (MK) as a gradual trend test and Pettitt's as an abrupt change test on annual WD and SL from 1950 to 2004 at nine hydrometer stations. Their study showed that long-term changes in annual WD, which were originally controlled by variation in precipitation, were not significant. SL at all main hydrometer stations showed declining trends during the study period. Declining trends were principally influenced by the construction of reservoirs and dams. In a mountainous tributary of the lower Xinjiang in China, hydrological response to changes in precipitation and anthropogenic activities were tested using MK and Pettitt's tests [11]. The power of MK and Spearman's rho tests to assess the significance of hydrological trends has been studied by Yue et al. [12]. They demonstrated that the power of both tests is directly proportional to trend slope, sample size, and predetermined significance level and inversely proportional to time series variation.

Ouyang et al. [6] established a relationship between soil erosion and landscape metrics at the Logliu catchment in China. They showed that landscape pattern significantly impacted soil erosion and sediment transportation. In several other studies, landscape metrics were applied at the landscape and patch levels to determine how hydrological conditions of the basin are affected by human activities such as land use change [13–17].

In recent decades, the Langat Basin has experienced rapid development towards urbanisation, industrialisation, and intense agriculture [18]. The Langat Basin is also a main source of drinking water for surrounding areas and a source of hydropower and has an important role in flood mitigation. Over the past four decades, the Langat Basin has served approximately 50% of the Selangor State population. However, the Selangor State is currently facing water shortage problems, especially in urban areas [19, 20].

This study was conducted to assess the impact of land use change, rainfall variation, and other anthropogenic manipulations on hydrological trends in selected upper catchments within the Langat Basin over a period of 25 years.

2. Methodology

2.1. Study area

The Langat Basin is located at the southern part of Klang Valley, which is the most urbanised river basin in Malaysia. It is believed that the Langat Basin is currently experiencing "spillover" effects due to the excessive development in the Klang Valley. Hydrometeorologically, the Langat Basin is affected by two types of monsoons, i.e., the Northeast (November–March) and the Southwest (May–September) [21]. The average annual rainfall is approximately 2400 mm. The wettest months are April and November, with an average monthly rainfall exceeding 250 mm, whereas the driest month is June, with an average monthly rainfall not exceeding 100 mm. Topographically, the Langat Basin can be divided into three distinct areas in reference to the Langat River, i.e., mountainous area in the upstream, undulating land in the centre, and flat flood plain in the downstream (**Figure 1**). The Langat Basin consists of a rich diversity of landform, surface feature, and land cover [21, 22].

Sub-basin	Lui	Hulu Langat	Semenyih
Main river	Lui	Langat	Semenyih
Geographic coordinate	3°07′–3°12′N and 101°52′–101°58′E	3°00′–3°17′N and 101°44′–101°58′E	2°55′–3°08′N and 101°49′–101°58′E
Drainage area (km^2)	68.25	390.26	235.62
Basin length (km)	11.5	34.5	26.5
Average slope (%)	35	29.4	27.4
Max. altitude (m)	1207	1479	1070
Min. altitude (m)	61	20	21
Ave. altitude (m)	354	277.4	243.9
Ref. hydrometer station	Sg. Lui	Sg. Langat	Sg. Semenyih
WD ($\times 10^6$ m^3/y)	55.05	289.64	146.11
SL ($\times 10^3$ ton/y)	5.88	146.6	36.81
Runoff ($mm/km^2/y$)	806.57	742.16	620.11
Sediment yield ($ton/km^2/y$)	86.22	375.65	156.22
Ref. rainfall station	Kg. Lui	UPM Serdang	Ldg. Dominion
Precipitation (mm)	2188.3	2453	2548.8
Land covers*	Forest 80.35%, cultivated rubber 9.85%, orchards 2.6%, mixed horticulture and crops, urbanised area, and mining land 7.2%	Forest 54.6%, cultivated rubber 15.6%, orchards 2%, urbanised area 15%, horticulture and crops, oil palm, lake, and mining land 12.8%	Forest 53.8%, cultivated rubber 17.4%, oil palm 6.3%, urban area 5.6%, secondary forest 3.6%, scrub land 2.4%, mining, other crops, mixed horticulture, orchard, bare land, marshland and aquaculture 10.9%

*Based on the land use map dated 2006.

Table 1. General information of the studied sub-basins.

Based on the availability of hydrometric stations in the Langat Basin, three sub-basin (upstream of the Langat River) were investigated. The descriptions about these sub-basin are given in **Table 1**.

2.2. Data set

WD, SL, and precipitation data between 1984 and 2008 recorded at all three hydrometer and rain gauge stations under study (**Table 1**) were obtained from the Department of Irrigation and Drainage (DID) of Malaysia. The geographic location and general information of the hydrometer stations are presented in **Figure 1** and **Table 2**. Land use maps dated 1984, 1988, 1990, 1995, 1997, 2001, 2002, 2005, and 2006 were obtained from the Soil Resource Management and Conservation Division, Department of Agriculture, Malaysia.

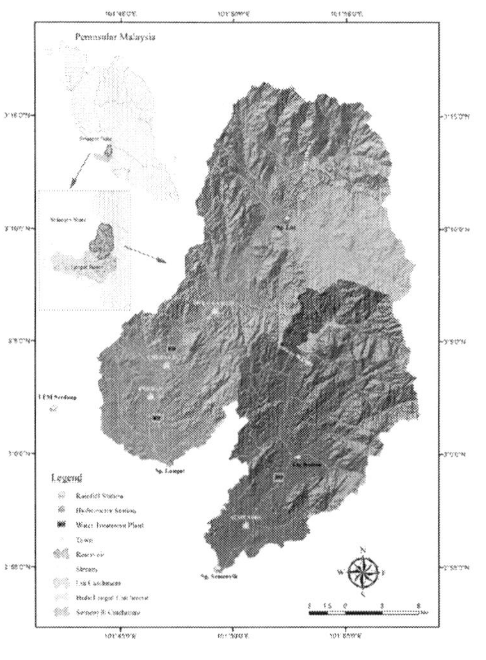

Figure 1. Geographic locations of the three study sub-basins.

2.3. Trend analysis

In this study, non-parametric tests, such as MK and Pettitt's, were used to detect gradual and abrupt changes in the hydrological data sets. According to Zhang et al. [4], non-parametric tests are preferred over parametric tests due to their strength in handling non-normally distributed data and missing data. The MK equation that is based on the *S* statistic is as follows:

http://dx.doi.org/10.5772/62463

$$S = \sum_{i=1}^{n-1} \sum_{j=i+1}^{n} sgn(x_j - x_i) \quad (1)$$

where *xi* and *xj* are sequential data values, *n* is the length of time series, and

$$\text{sgn}(\theta) = \begin{cases} 1 \ \ if\ \theta > 0 \\ 0 \ \ if\ \theta = 0 \\ -1 \ \ if\ \theta < 1 \end{cases} \quad (2)$$

Mann (1945) and Kendall (1975) (as cited by Yue et al. [12]) have posted that, when $n \geq 8$, S is almost normally distributed with the following mean and variance:

$$E(S) = 0 \quad (3)$$

$$V(S) = \frac{n(n-1)(2n+5) - \sum_{i=1}^{n} t_i i(i-1)(2i+5)}{18} \quad (4)$$

where t_i is number of ties of the extent i.

The standard Z statistic is calculated as follows:

$$Z_{MK} = \begin{cases} \frac{S-1}{\sqrt{Var(S)}} & S > 0 \\ 0 & S = 0 \\ \frac{S+1}{\sqrt{Var(S)}} & S < 0 \end{cases} \quad (5)$$

Z_{MK} pursues the standard normal distribution with μ=0 and δ=1.

The probability (P) of the S statistic is estimated by normal cumulative distribution function as follows:

$$P = 0.5 - \Phi(|Z|)$$

$$\Phi(|Z|) = \frac{1}{\sqrt{2\pi}} \int_0^{|Z|} e^{\frac{-t^2}{2}} dt \quad (6)$$

Statistical significance of the data trend was based on 95% confidence level [12].

The MK test is not robust against autocorrelation [4, 11, 23]. As such, an autocorrelation test was performed on the data set to determine the degree of autocorrelation. The autocorrelation coefficient is estimated as follows [24]:

$$r_k = \frac{\frac{1}{n-k}\sum_{t=1}^{n-k}(X_t - \bar{X})(X_{t+k} - \bar{X})}{\frac{1}{n-1}\sum_{t-1}^{n}(X_t - \bar{X})} \tag{7}$$

where $X=t=1nXtn$, n is the sample size, and k is the lag.

For a completely random series, $r_k \approx 0$ for all $k \neq 0$. If a series of r_k (for $k \neq 0$) fall between the 95% confidence level estimated by $\frac{u}{l} = \left(-1 \pm Z_{1-\propto/2}\sqrt{n-2}\right)/(n-1)$ (where n is the length of tested time series, l and u are the lower and upper limits, α is the significance level, and Z is the critical value of standard normal distribution for a given α), then the tested series will be independent at the 95% confidence level [23]. The WD and SL data showed significant autocorrelation. Therefore, Zhang's method of data pre-whitening [25] was used to eliminate significant autocorrelation within the data.

Sen's non-parametric method was used to estimate the change magnitude (i.e., slope of the linear trend). Sen's method is robust against non-normally distributed data, missing values, and extreme outliers (Sen, 1968) (as cited by Zhang and Lu [11]).

Considering a sequence of random variables $X_1, X_2, \ldots, X_t$, which have a change point at τ [X_t for $t = 1, 2, \ldots, \tau$ have a common distribution function $F_1(x)$ and X_t for $t=\tau+1, \ldots, T$ have a common distribution function $F_2(x)$ and $F_1(x) \neq F_2(x)$], Pettitt's test (1979) (as cited by Zhang et al. [4] and Wolfe and Schechtman [26]) was used to detect one unknown change point in the pre-whitened WD and SL time series. In the Pettitt's test, null hypothesis (H_0): no change ($\tau=T$) is tested against alternative hypothesis (H_a): change ($1 \leq \tau < T$)) by the non-parametric K statistic, as follows:

$$K_t = max_{1 \leq t \leq T}\left|U_{t,T}\right| = \max\left(K_T^+, K_T^-\right) \tag{8}$$

where $U_{t,T} = \sum_{i=1}^{t}\sum_{j=t+1}^{T} \mathrm{sgn}(X_t - X_j)$, $\mathrm{sgn}(\theta) = \begin{cases} 1\,(if\ \theta > 0) \\ 0\,(if\ \theta = 0) \\ -1\,(if\ \theta < 0) \end{cases}$. $K_T^+ = max_{1 \leq t \leq T} U_{t,T}$ for downward shift and $K_T^- = -min_{1 \leq t \leq T} U_{t,T}$ for upward shift. The significance level of K_T^+ or K_T^- is estimated by $P = \exp\left(\frac{-6K_T^2}{T^3 + T^2}\right)$. When K_T occurs, the time t will be the point of change. When P is smaller than the specific significance level, H_0 is rejected.

The above procedures were performed using XLSTAT and R statistical packages.

2.4. Landscape analysis

To assess the changes in land use patterns over the period 1984–2006 (including nine records), Patch Analyst 3.0 (Grid) extension in ArcView 3.3 was applied to calculate landscape metrics [27], which are fundamental indices for the detection of trends in land use change [6]. A brief description of the six selected landscape metrics for this work is given below [28, 29]:

The number of patches (NUMP) is ≥1.

The patch size coefficient of variation (PSCOV) is the variability of the patch size relative to the mean patch size.

$$PSCOV = \frac{PSSD}{MPS}(100) \tag{9}$$

where PSCOV ≥ 0, PSSD is the standard deviation in patch size and MPS is mean patch size of the corresponding patch type.

Edge density (ED) equals the sum of lengths (m) of all edge segments involving the corresponding patch type divided by the total landscape area (m^2) in meters per hectare.

The Shannon's diversity index (SDI; at the landscape level) is a measure between 0 and 1. SDI equals 0 when the landscape comprises only one patch (i.e., no diversity) and increases with increasing number of patch types. SDI equals 1 when the different patch types are distributed proportionally.

$$SDI = -\sum_{i=1}^{m}(P_i.lnP_i) \tag{10}$$

where P_i is proportion of the landscape occupied by the patch type (class) i.

The Shannon's evenness index (SEI; at the landscape level) is a measure between 0 and 1. SEI equals 0 when the landscape comprises only one patch (i.e., no diversity) and approaches 0 as the areal distribution of patch types becomes uneven (i.e., dominated by one type). When the areal distribution of patch types is perfectly even, SEI equals 1.

$$SEI = \frac{-\sum_{i=1}^{m}(P_i.lnP_i)}{lnm} \tag{11}$$

where m defines the number of patch types (classes) present in the landscape including the landscape border.

The interspersion and juxtaposition index (IJI) is the observed interspersion over the maximum possible interspersion for a given number of patch types. IJI approaches 0 when the corresponding patch type is adjacent to only one other patch type and is 100 when the corresponding patch type is equally adjacent to all other patch types.

$$IJI = \frac{-\sum_{k=1}^{m'}\left[\left(\sum_{k=1}^{m'}\frac{e_{ik}}{e_{ik}}\right) ln\left(\sum_{k=1}^{m'}\frac{e_{ik}}{e_{ik}}\right)\right]}{\ln(m'-1)}(100) \quad (12)$$

where e_{ik} is the total length (m) of edge in landscape between the patch types (classes) i and k, and m' defines the number of patch types (classes) presented in the landscape.

3. Results

3.1. Hydrological trend analysis

The autocorrelation test reveals that the WD series (except that at Sg. Lui hydrometer station) and SL series have at least one autocorrelation coefficient that is significant at the 95% confidence level (**Figure 2**). The autocorrelation coefficients of the WD series at both Sg. Langat

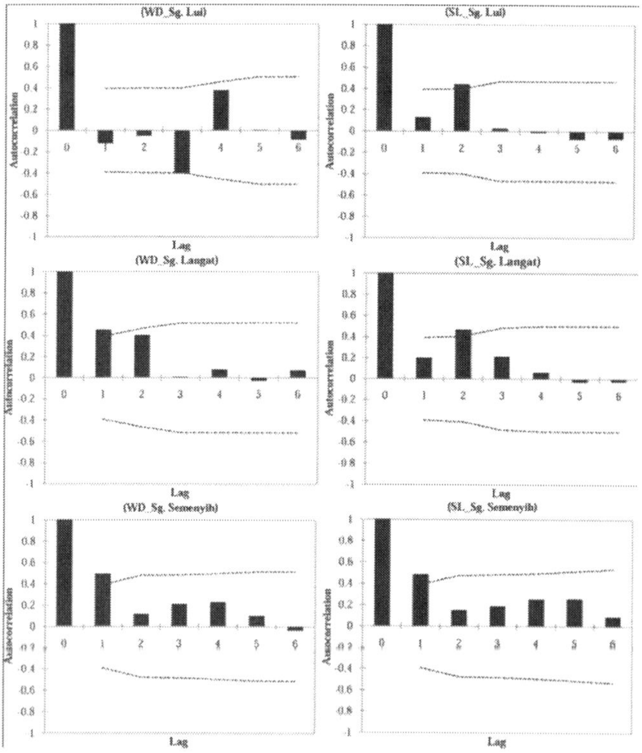

Figure 2. Autocorrelograms, resulted from autocorrelation test on WD and SL at the selected hydrometer stations.

and Sg. Semenyih hydrometer stations are significant at the first lag. The SL series at Sg. Langat and Sg. Semenyih hydrometer stations have one significant autocorrelation coefficient at the second and first lags, respectively. The SL series at Sg. Lui hydrometer station is autocorrelated significantly only at the second lag.

The results of gradual trend analysis based on MK and Pre-Whitening MK (PWMK) tests for WD and SL data are shown in **Table 2**. At Sg. Langat, WD shows an increasing trend that is significant at the 95% confidence level; however, the ascendant trend of SL is not statistically significant. There are no significant trends in the hydrological time series of Sg. Lui and Sg. Semenyih.

Gradual changes in the hydrological time series at all three hydrometer stations are given in **Table 2**. At Sg. Lui, WD and SL are decreasing at a rate of $0.524{\times}10^6$ m^3/y and $0.119{\times}10^3$ ton/y, respectively. At Sg. Langat, however, WD and SL are increasing at a rate of $9.899{\times}10^6$ m^3/y and $1.415{\times}10^3$ ton/y. At Sg. Semenyih, WD and SL show declining tendencies at $3.686{\times}10^6$ m^3/y and $0.316{\times}10^3$ ton/yr.

Station name	Parameter	MK and PWMK trend tests			Sen's slope estimator			
		τ	P	Trend	Trend	Trend_P	Linear	Intercept
Sg. Lui	WD	−0.153	0.297	Decreasing	−0.524	−13.105	−0.525	62.723
	SL	−0.072	0.637	Decreasing	−0.119	−2.986	−0.471	4.596
Sg. Langat	**WD**	**0.326**	**0.027**	**Increasing**	**9.899**	**247.485**	**11.531**	**156.869**
	SL	0.130	0.385	Increasing	1.415	35.380	13.142	46.278
Sg. Semenyih	WD	−0.196	0.189	Decreasing	−3.686	−92.145	−3.912	187.036
	SL	−0.058	0.710	Decreasing	−0.316	−7.909	−3.611	20.707

Trend: Sen's slope (trend) per unit time; Trend_P: Sen's slope (trend) over the time period; Linear: least-squares fit trend; Intercept: intercept of the Sen's slope (trend).

Table 2. Results of MK and PWMK tests with the Sen's slope estimator (at α=0.05) applied on WD and SL (data in bold are significant).

The results of abrupt changes based on the Pettitt's test for WD and SL are shown in **Figures 3** and **4** and **Table 3**. The results show significant drastic changes in the hydrological time series at Sg. Langat and Sg. Semenyih hydrometer stations. At Sg. Langat, the mean level of WD (after 1998) shifted upward to $392.09{\times}10^6$ m^3/y, which corresponds to a 77% increase, whereas the mean level of SL (after 1999) shifted upward to $297.27{\times}10^3$ ton/y, which corresponds to a 380% increase (**Figure 3**). At Sg. Semenyih, the mean level of WD (after 1993) shifted downward to $111.18{\times}10^6$ m^3/y (44% decrease) and the mean level of SL shifted downward to $14.89{\times}10^3$ ton/y (78% decrease; **Figure 4**). At Sg. Lui, however, downward shifts in hydrological time series are not significant.

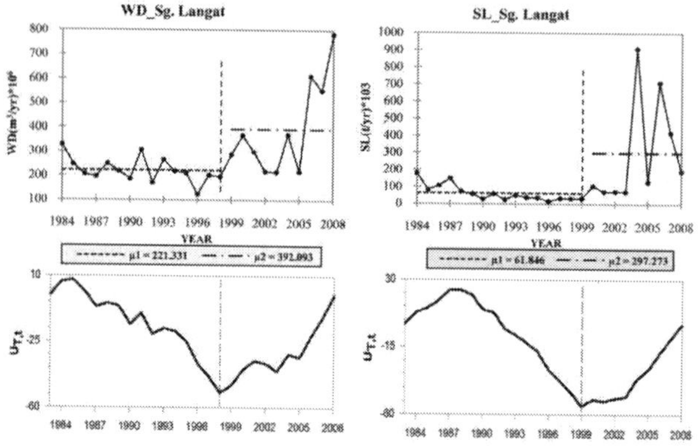

Figure 3. Abrupt changes in the mean level of WD and SL for Sg. Langat at the significance level of 0.05.

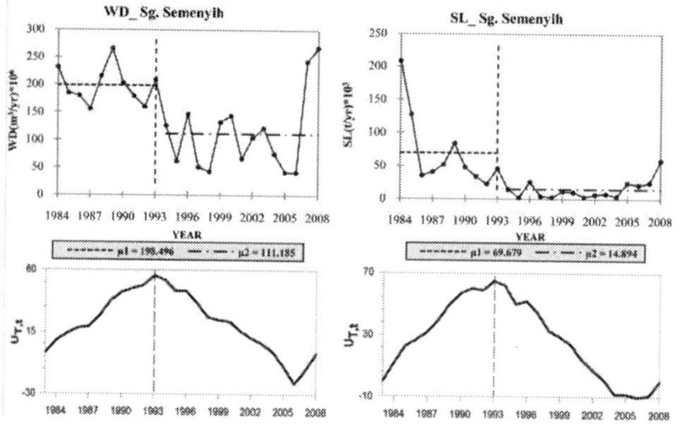

Figure 4. Abrupt changes in the mean level of WD and SL for Sg. Semenyih at the significance level of 0.05.

Station name	Parameter	K_T	*P*	Shift	*T*
Sg. Lui	Water	56.000	0.430	Downward	1987
	Sediment	82.000	0.084	Downward	1988
Sg. Langat	**Water**	**104.000**	**0.012**	**Upward**	**1998**
	Sediment	**108.000**	**0.009**	**Upward**	**1999**
Sg. Semenyih	**Water**	**112.000**	**0.010**	**Downward**	**1993**
	Sediment	**130.000**	**0.005**	**Downward**	**1993**

Table 3. Results of Pettitt's test applied on WD and SL (data in bold are significant at the level of 0.05).

http://dx.doi.org/10.5772/62463

Based on the PWMK test, the gradual change in the Lui and Semenyih sub-basin are showing declining tendencies. It is important to know whether these declining tendencies will prevail after the change point. **Table 4** shows that all hydrological series have increasing tendencies after the change point, which are statistically significant only for SL series at Sg. Semenyih. Meanwhile, decreasing tendencies before the change point are statistically significant only for SL series recorded at Sg. Langat.

Station	Parameter	*T*	Time series	MK trend test		
				τ	*P*	Trend
Sg. Lui	WD	1987	Pre-T[*]	–	–	–
			Post-T	0.057	0.739	Increasing
	SL	1988	Pre-T	−0.666	0.308	Decreasing
			Post-T	0.184	0.269	Increasing
Sg. Langat	WD	1998	Pre-T	−0.331	0.112	Decreasing
			Post-T	0.405	0.127	Increasing
	SL	1999	**Pre-T**	**−0.582**	**0.004**	**Decreasing**
			Post-T	0.388	0.175	Increasing
Sg. Semenyih	WD	1993	Pre-T	0.000	1.000	–
			Post-T	0.143	0.511	Increasing
	SL	1993	Pre-T	−0.357	0.265	Decreasing
			Post-T	**0.450**	**0.028**	**Increasing**

[*]Limitation in number of records.

Table 4. Results of the MK and PWMK tests on WD and SL before and after the change points (data in bold are significant at the level of 0.05).

3.2. Landscape analysis

The relationships between landscape metrics and hydrological variables are given in **Table 5**. At the Lui sub-basin, all metrics with the exception of IJI and SEI are negatively correlated with WD and SL. SL correlates significantly with PSCOV and SEI, whereas WD correlates significantly with ED and NUMP.

At the Semenyih sub-basin, all metrics are negatively correlated with WD and SL. Correlations between these metrics and WD are statistically significant. Correlations between these metrics with the exception of IJI and PSCOV and SL are statistically significant. At the Hulu Langat sub-basin, on the contrary, all metrics are positively correlated with WD and SL but are not statistically significant.

In comparison, correlations between hydrological series and landscape metrics are more pronounced after the change point (**Table 5**). For example, at the Semenyih sub-basin, correlations between hydrological series and all metrics change from negative to positive and are only significant for PSCOV. Also, at the Hulu Langat sub-basin, correlations between WD and the metrics ED, SDI, and NUMP are statistically significant.

Table 6 shows the trend analysis of landscape metrics during the period 1984 to 2006. At the Lui **sub**-basin, only NUMP shows a significant increasing trend. At the Hulu Langat **sub**-basin, all metrics show increasing tendencies but are statistically significant only for PSCOV, ED, SDI, and NUMP. At the Semenyih **sub**-basin, all metrics with the exception of IJI show significant increasing trends.

Landscape metric	All records						Records after the change points					
	Lui		Hulu Langat		Semenyih		Lui		Hulu Langat		Semenyih	
	WD	SL	WD	SL	WD	SL	WD	SL	WD	SL	WD	SL
PSCOV	−0.625	**−0.788***	0.358	0.243	**−0.921****	−0.584	−0.408	0.081	0.460	0.674	**0.887***	**0.932****
ED	**−0.680***	−0.471	0.289	0.244	**−0.957****	**−0.790***	−0.549	−0.026	**0.941***	0.784	0.552	0.548
IJI	0.091	0.238	0.206	0.191	**−0.973****	−0.660	0.048	0.137	0.674	0.659	0.380	0.575
SDI	−0.556	−0.099	0.257	0.228	**−0.885****	**−0.865****	−0.532	−0.026	**0.981***	0.846	0.789	0.670
SEI	0.474	**0.951****	0.340	0.379	**−0.940****	**−0.873****	−0.150	0.168	0.481	0.646	0.389	0.398
NUMP	**−0.703***	−0.481	0.362	0.523	**−0.954****	**−0.704***	−0.584	−0.025	**0.955***	0.870	0.762	0.759

*Correlation is significant at the 0.05 level.
**Correlation is significant at the 0.01 level.

Table 5. Correlations between the different landscape metrics and hydrological series using the Pearson correlation method.

Landscape metric	Lui	Hulu Langat	Semenyih
PSCOV	NS ↑	* ↑	* ↑
ED	NS ↑	* ↑	* ↑
IJI	NS ↓	NS ↑	NS ↑
SDI	NS ↑	* ↑	* ↑
SEI	NS ↓	NS ↑	* ↑
NUMP	* ↑	* ↑	* ↑

*Trend is significant at the 0.05 level.
NS, not significant. ↑, increasing; ↓, decreasing.

Table 6. Trend analysis of the landscape metrics during 1984 to 2006 at the studied sub-basins.

The categorisation of the landscape metrics using the clustering technique for all three sub-basins is given in **Figures 5–7**.

At the Lui **sub**-basin, PSCOV, ED, and NUMP corresponding to years 1984, 1988, and 1990 are categorised in the first cluster, whereas landscape metric values of the remaining years are classified in the second cluster. Meanwhile, SEI (1984 and 1988) and SDI (1988 and 1990) are categorised in the first cluster, whereas the values of the remaining years are classified in the second cluster (**Figure 5**).

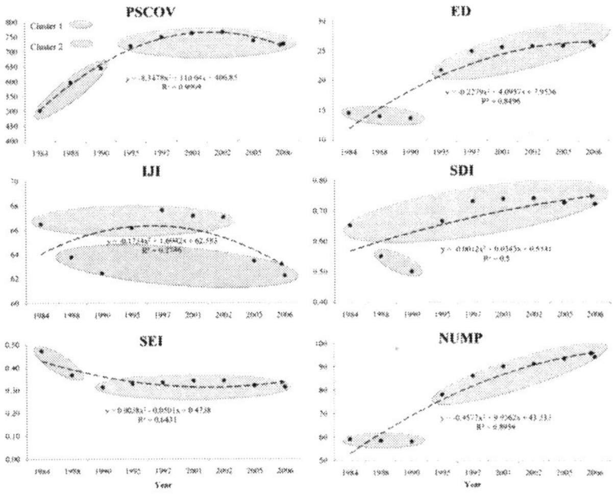

Figure 5. Change trends and classification of the landscape metrics at the Lui sub-basin during 1984 to 2006.

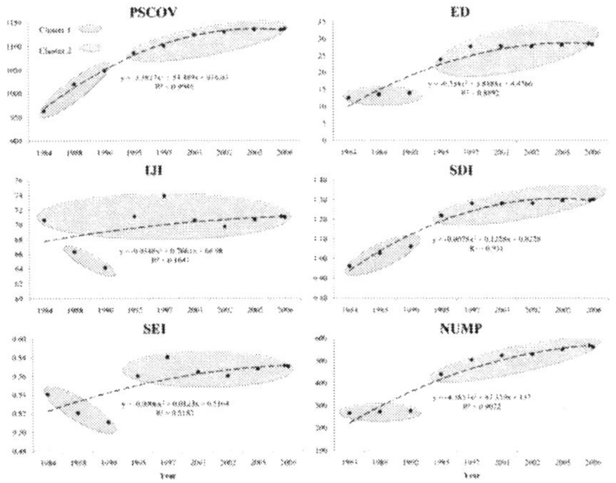

Figure 6. Change trends and classification of the landscape metrics at the Hulu Langat Sub Basin during 1984-2006.

All landscape metric values at the Hulu Langat (except for IJI) and Semenyih sub-basins corresponding to years 1984, 1988, and 1990 are grouped in the first cluster and the values of the remaining years are grouped in the second cluster (**Figure 6**).

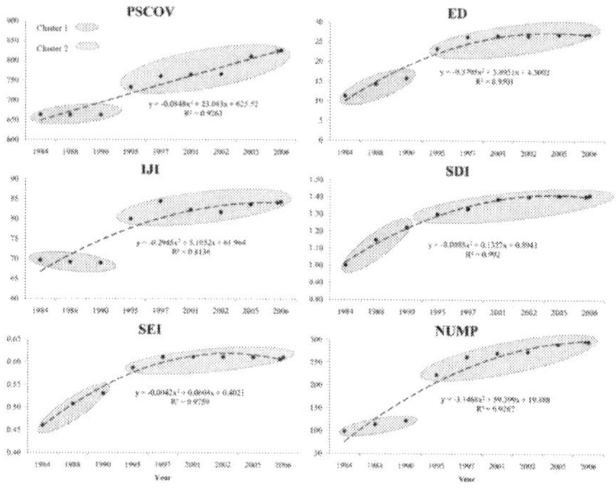

Figure 7. Change trends and classification of the landscape metrics at the Semenyih sub-basin during 1984 to 2006.

3.3. Land use change detection

Change detection was based on the land use maps dated 1984 and 2006. The results show noticeable gains in agriculture, bare land, mining, oil palm, and urban acreage and a remarkable loss in rubber acreage (**Table 7**).

Land use	Difference (2006–1984) in hectares		
	Lui	Hulu Langat	Semenyih
Agriculture	436.39	1545.32	451.57
Bare land	6.30	370.60	175.50
Forest	359.75	112.04	-237.72
Grassland	5.00	–471.70	382.30
Marshland/swamp	46.42	148.93	-39.73
Mining	16.67	395.27	147.57
Oil palm	3.44	21.81	1051.89
Rubber	–809.43	–7817.72	-3860.35
Urban/built-up area	111.09	5649.90	1501.61
Water body	0.00	45.55	427.36

Table 7. Land use change detection between 1984 and 2006 at the studied sub-basins.

http://dx.doi.org/10.5772/62463

Increasing gradual trends (as determined by the MK test using the 1984–2006 time series) in agriculture and mining acreages are significant only at the Hulu Langat **sub**-basin, whereas increasing gradual trend in oil palm acreage is significant only at the Semenyih **sub**-basin. Decreasing gradual trend in rubber acreage is significant at both Hulu Langat and Semenyih sub-basins. At all three sub-basins, a change trend in forest area is not significant, whereas increasing trend in the urbanised area is significant (**Table 8**).

Land use	Lui	Hulu Langat	Semenyih
Agriculture	NS ↑	* ↑	NS ↑
Forest	NS ↓	NS ↓	NS ↓
Urban	* ↑	* ↑	* ↑
Rubber	NS ↓	* ↓	* ↓
Oil palm	NS ↑	NS ↑	* ↑
Mining	NS ↑	* ↑	NS ↑

*Trend is significant at the 0.05 level.
NS, not significant. ↑, increasing; ↓, decreasing.

Table 8. Trend analysis of land use change during 1984–2006 at the studied sub-basins.

4. Discussion

Based on the hydrological trend analysis, the Hulu Langat sub-basin showed a significant increasing trend of WD. However, the Semenyih and Lui sub-basins showed decreasing tendencies of WD and SL. Gradual increase in hydrological series after the change points is significant only in the case of SL at Sg. Semenyih. In the following sections, hydrological alterations are discussed in relation to land use change, rainfall fluctuations, and other anthropogenic manipulations.

4.1. Effect of land use/cover change (LUCC)

Based on the data from the Department of Statistics, Malaysia (2001) and the National Urbanisation Policy of Malaysia (1981), rapid development in the state of Selangor started in 1981. The rapid development was aimed at attracting approximately 18% of Malaysia's population to be settled in the state of Selangor by the year 2000 [30, 31]. The Langat Basin appears as a suitable barometer to measure urbanisation and agricultural/industrial development in the state of Selangor.

Based on **Tables 7** and **9**, rubber acreage at the Hulu Langat and Semenyih sub-basins decreased significantly between 1984 and 2006. During this period, at the Hulu Langat sub-basin, 34% of rubber acreage was transformed into urban areas, whereas another 11% was used for other agricultural production. Similarly, at the Semenyih sub-basin, 21% of rubber acreage was transformed into urban areas, whereas 15% and 6% were used for oil palm and other agricultural productions, respectively (**Table 7**). Based on these results and the work of Noorazuan et al. [22] and Juahir et al. [19], it is expected that these changes will affect the stream flow behaviour and characteristics.

1984	**2006**						
	Lui						
	Forest	**Rubber**	**Agriculture**	**Scrub/idle grassland**	**Oil palm**	**Urbanised/ industrial area**	**Water bodies**
Forest		1	1			1	
Rubber			21			5	
Scrub/idle grassland	47	11	33			4	
	Hulu Langat						
Forest		4				3.5	
Rubber			11			34	
Agriculture		17				26	
Scrub/idle grassland	38	17	10			31	
	Semenyih						
Forest		2.5				1.4	2.7
Rubber			6		15	21	
Agriculture		7		10	56	15	
Scrub/idle grassland	70	10				19	
Oil palm		8	8			34	
Swamp/ marshland		7	80		13		
Bare land		67					

Table 9. Land use change matrix for important transitions (frequencies in %) between the years 1984 and 2006 at the studied sub-basins.

Tables 6 shows the increasing trend in landscape change during 1984–2006, which is confirmed by correlation analysis (**Table 5**), especially after the change points [6, 32].

At the Hulu Langat sub-basin, cluster analysis shows that discriminant points between the clusters of landscape metrics (except for IJI) are within the period 1990–1997 (**Figure 6**). These points correspond to the points of change in WD and SL. At the Semenyih sub-basin, the change point in WD and SL (i.e., 1993) matches the discriminant point in all landscape metric clusters, with the exception of SDI (**Figure 7**). At the Lui sub-basin, points of change in WD and SL (1987 and 1988) match only the discriminant point in SEI. As indicated in **Table 6**, at the Lui sub-basin, change trends in landscape metrics are not statistically significant. This could have contributed to the insignificant impact of LUCC on the basin hydrological conditions.

4.2. Effect of rainfall variations

The rainfall stations (i.e., Kg. Lui, UPM Serdang, and Ldg. Dominion) were analysed for rainfall change trend. These three rainfall stations represent rainfall events at the Lui, Hulu Langat, and Semenyih sub-basins, respectively (**Table 10**). Increasing trend in the rainfall time series is only significant at UPM Serdang, which corresponds to increasing trends in WD and SL at Sg. Langat. Although the change point in rainfall series at UPM Serdang (i.e., 1998) is not statistically significant, it matches the change point in WD and SL at Sg. Langat. This point matches the critical water level at Langat Reservoir [33], which has been reported by Shaaban and Low [34]. At Kg. Lui, the change tendency of rainfall series after the hydrological change point matches the trend of WD and SL at Sg. Lui (**Table 11**). At Ldg. Dominion, the change tendency in rainfall does not match the available tendency in hydrological series at Sg. Semenyih, especially for SL after the change point.

Station name	**MK trend test**			**Pettitt's test**			
	τ	P	**Trend**	K_T	P	**Shift**	T
Kg. Lui	0.144	0.333	Increasing	70.000	0.188	Upward	2003
UPM Serdang	**0.341**	**0.020**	**Increasing**	78.000	0.079	Upward	1998
Ldg. Dominion	0.101	0.503	Increasing	68.000	0.231	Upward	1990

Table 10. Results of the PWMK and Pettitt's tests applied on the rainfall time series at the representative stations (data in bold are significant at α=0.05).

Rainfall station	**T_Hydro series**	**Time series**	**PWMK trend test**		
			τ	P	**Trend**
Kg. Lui	1988	Pre-T	−0.333	0.734	Decreasing
		Post-T	0.111	0.528	Increasing
UPM Serdang	1998	Pre-T	0.256	0.246	Increasing
		Post-T	0.066	0.858	Increasing
Ldg. Dominion	1993	Pre-T	0.444	0.117	Increasing
		Post-T	0.000	1.000	–

Table 11. Results of PWMK test applied on the rainfall time series before and after the hydrological change points.

Thus far, results reveal the significant impacts of land use and rainfall variations on WD at the Hulu Langat sub-basin. However, the impacts on the sub-basin SL are not statistically significant. At the Lui sub-basin, the change trends in rainfall and landscape variables are not statistically significant. Hence, the insignificance of hydrological series trend is expected. At the Semenyih sub-basin, the impact of land use change on hydrological series is driven by the significant increasing trend in SL after 1993. However, rainfall variations do not impact the trend of hydrological series.

From the preceding discussion, two questions are important. First, have the Semenyih Reservoir and its connected water treatment facilities at the Semenyih sub-basin impacted the basin WD significantly? Secondly, despite the significant impact of land use change on the change trend in WD at the Hulu Langat sub-basin, why is the change trend in SL not statistically significant? In the following discussion, these questions are addressed.

4.3. Effect of man-made structures

There are two strategic dams in the Langat Basin. The Langat Dam, constructed in 1979, has a drainage catchment area of 41.5 km^2 and a reservoir capacity of 37.5 Mm^3. The Semenyih Dam, built in 1985, has a drainage catchment area of 56.7 km^2 and a reservoir capacity of 62.6 Mm^3. Both these dams supply domestic and industrial water. The Langat Dam is also used to generate power supply at moderate capacity for consumption within the Langat Valley. Currently, there are three major water treatment plants (WTP; operating 24 hours a day) within the study area. The Sg. Langat and Cheras WTPs along the Langat River produce 386.4 and 27 million litres per day (MLD) of clean water, respectively. The Semenyih WTP along the Semenyih River produces 545 MLD of clean water [30, 33].

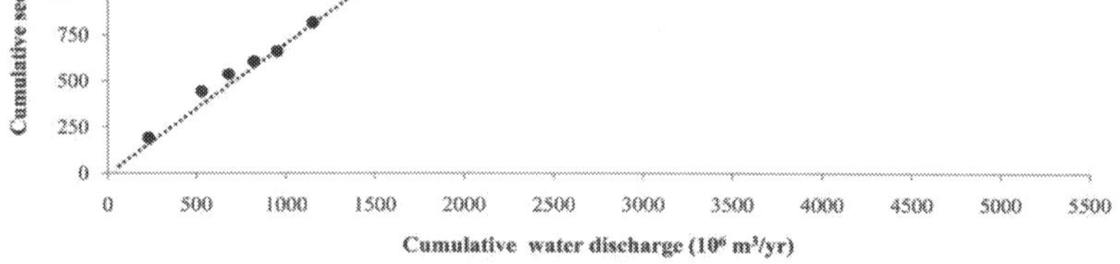

Figure 8. Cumulative double mass plot at Sg. Semenyih.

To evaluate the impact of Semenyih Dam construction on trends of hydrological time series, a double mass curve was plotted. As illustrated in **Figure 8**, at Sg. Semenyih, the hydrological time series trend after 1985 is seriously affected by the dam and WTP construction. The mean WD level changes from 231.8×10^6 m^3/y before 1985 to 141×10^6 m^3/y after 1985. The results of the Pettitt's test for WD and SL during the period 1975–1993 were statistically significant (P=0.018 and P<0.001, respectively). This confirms 1985 as the point of change during the period 1975 to 1993.

Shaaban and Low [34] showed that drought events reduced WD at the Semenyih sub-basin, particularly in the period 1993–1998. As such, WTP and dam together with the effect of drought have been able to reduce the increasing trend of WD, especially after 1993.

At the Hulu Langat sub-basin, due to the significant trend in urbanisation and agricultural activities (**Tables 7** and **9**), the number and size of natural or artificial ponds are expected to increase dramatically. Field observation from this study confirms that the quantity of natural and artificial ponds is higher at Hulu Langat compared to that at Semenyih. The ponds are believed to affect the sedimentation process by increasing the deposition rate, hence resulting in the reduction in SL of the basin (**Figure 9**). It is clear that the Langat Dam and other sediment trapping features (i.e., natural and artificial ponds) contributed to the insignificant trend of SL at the Hulu Langat sub-basin.

Figure 9. Ponds arisen from urban and agricultural development at the Hulu Langat sub-basin (extracted from SPOT 5 satellite images, dated 2006).

5. Conclusion

Increasing trend in WD at the Hulu Langat sub-basin was originally controlled by significant variations in land use and rainfall. However, increasing trend in SL was not significant due to dam construction and increase in the number and size of sediment trapping features, which is due to urbanisation and agricultural activities. At the Semenyih sub-basin, increasing trend in SL after 1993 was closely related to significant trends in landscape metrics and land use changes. However, WD did not increase significantly after 1993, primarily due to the impact of dam and water treatment facilities and continuous drought until 1998. At the Lui sub-basin, trends in land use and rainfall variations were mostly insignificant, thus causing an insignificant change in hydrological series.

This study demonstrates the power of the PWMK and Pettitt's tests for trend evaluation of hydrological time series. The results obtained in this work are consistent with studies done by other researchers [35–45]. Also, integrating landscape analysis with statistical analyses as emphasised in the work of several others [6, 13–17, 32, 46, 47] could increase the depth of interpretation with regard to the complex hydrological conditions of developed basins.

Acknowledgements

The authors acknowledge Universiti Putra Malaysia for procuring land use maps and satellite imagery and the Department of Irrigation and Drainage, Malaysia, for supplying hydrological data.

Author details

Hadi Memarian[1*] and Siva K. Balasundram[2]

*Address all correspondence to: hadi_memarian@birjand.ac.ir

1 Assistant Professor, Department of Watershed Management, Faculty of Natural Resources and Environment, University of Birjand, Birjand, Iran

2 Associate Professor, Department of Agriculture Technology, Faculty of Agriculture, Universiti Putra Malaysia (UPM), Serdang, Malaysia

References

[1] Dai, Z., Liu, J. T., & Xiang, Y. (2015). Human interference in the water discharge of the Changjiang (Yangtze River), China. *Hydrological Sciences Journal*, 60(10), 1770–1782.

[2] Zhang X., Cao W., Guo Q., & Wu S. (2010) Effects of land use change on surface runoff and sediment yield at different watershed scales on the Loess Plateau. *International Journal of Sediment Research*, 25, 283–293.

[3] García-Ruiz J. M., Regüés D., Alvera B., Lana-Renault N., Serrano-Muela P., Nadal-Romero E., et al. (2008) Flood generation and sediment transport in experimental catchments affected by land use changes in the central Pyrenees. *Journal of Hydrology*, 356, 245–260.

[4] Zhang S., Lu X. X., Higgitt D. L., Chen C. T. A., Han J., & Sun H. (2008) Recent changes of water discharge and sediment load in the Zhujiang (Pearl River) Basin, China. *Global and Planetary Change*, 60, 365–380.

[5] Shi Z. H., Huang X. D., Ai L., Fang N. F., & Wu G. L. (2014). Quantitative analysis of factors controlling sediment yield in mountainous watersheds. *Geomorphology*, 226, 193–201.

[6] Ouyang W., Skidmore A. K., Hao F., & Wang T. (2010) Soil erosion dynamics response to landscape pattern. *Science of the Total Environment*, 408, 1358–1366.

[7] Ghaffari, G., Keesstra, S., Ghodousi, J., & Ahmadi, H. (2010). SWAT-simulated hydrological impact of land-use change in the Zanjanrood Basin, Northwest Iran. *Hydrological processes*, 24(7), 892–903.

[8] Li Z., Liu W. Z., Zhang X. C., & Zheng F. L. (2009) Impacts of land use change and climate variability on hydrology in an agricultural catchment on the Loess Plateau of China. *Journal of Hydrology*, 377, 35–42.

[9] He H., Zhou J., & Zhang W. (2008) Modelling the impacts of environmental changes on hydrological regimes in the Hei River Watershed, China. *Global and Planetary Change*, 61, 175–193.

[10] Nearing M. A., Jetten V., Baffaut C., Cerdan O., Couturier A., Hernandez M., et al. (2005) Modeling response of soil erosion and runoff to changes in precipitation and cover. *CATENA*, 61, 131–154.

[11] Zhang S. & Lu X. X. (2009) Hydrological responses to precipitation variation and diverse human activities in a mountainous tributary of the lower Xijiang, China. *Catena*, 77, 130–142.

[12] Yue S., Pilon P., & Cavadias G. (2002) Power of the Mann-Kendall and Spearman's ϱ tests for detecting monotonic trends in hydrological series. *Journal of Hydrology*, 259, 254–271.

[13] Lin Y. P., Hong N. M., Wu P. J., Wu C. F., & Verburg P. H. (2007) Impacts of land use change scenarios on hydrology and land use patterns in the Wu-Tu watershed in northern Taiwan. *Landscape and Urban Planning*, 80, 111–126.

[14] Weng Y. C. (2007) Spatiotemporal changes of landscape pattern in response to urbanization. *Landscape and Urban Planning*, 81, 341–353.

[15] Song I. J., Hong S. K., Kim H. O., Byun B., & Gin Y. (2005) The pattern of landscape patches and invasion of naturalized plants in developed areas of urban Seoul. *Landscape and Urban Planning*, 70, 205–219.

[16] Cifaldi R. L., David Allan J., Duh J. D., & Brown D. G. (2004) Spatial patterns in land cover of exurbanizing watersheds in Southeastern Michigan. *Landscape and Urban Planning*, 66, 107–123.

[17] Crews-Meyer K. A. (2004) Agricultural landscape change and stability in northeast Thailand: historical patch-level analysis. *Agriculture, Ecosystems & Environment*, 101, 155–169.

[18] Mohamed A. F., Yaacob W. Z., Taha M. R., & Samsudin A. R. (2009) Groundwater and soil vulnerability in the Langat Basin Malaysia. *European Journal of Scientific Research*, 27, 628–635.

[19] Juahir H., Zain S. M., Yusoff M. K., Hanidza T. I. T., Armi A. S. M., Toriman M. E., & Mokhtar M. (2010) Spatial water quality assessment of Langat River Basin (Malaysia) using environmetric techniques. *Environmental Monitoring and Assessment*, published online at Springerlink.com.

[20] Ayub K. R., Hin L. S., & Aziz H. A. (2009) SWAT application for hydrologic and water quality modeling for suspended sediments: a case study of Sungai Langat's Catchment in Selangor. *International Conference on Water Resources (ICWR 2009)*, Langkawi, Kedah, Malaysia, published online at redac.eng.usm.my.

[21] Memarian H., Balasundram S. K., Talib J., Teh C. B. S., Alias M. S., Abbaspour K. C., & Haghizadeh A. (2012) Hydrologic analysis of a tropical watershed using KINEROS2. *Environment Asia*, 5, 84–93.

[22] Noorazuan M. H., Rainis R., Juahir H., Zain S. M., & Jaafar N. (2003) GIS application in evaluating land use-land cover change and its impact on hydrological regime in Langat River Basin, Malaysia. *Conference of MapAsia 2003*, Malaysia, Kuala Lumpur, published online at www.geospatialworld.net.

[23] Zhang Q., Xu C. Y., Becker S., & Jiang T. (2006) Sediment and runoff changes in the Yangtze River Basin during past 50 years. *Journal of Hydrology*, 331, 511–523.

[24] Haan C. T. (2002) Statistical Methods in Hydrology. 2nd edn. Ames, Iowa State Press, 496 pp.

[25] Wang X. L. & Swail V. R. (2001) Changes of extreme wave heights in Northern Hemisphere oceans and related atmospheric circulation regimes. *Journal of Climate*, 14, 2204–2221.

[26] Wolfe D. A. &; Schechtman E. (1984) Nonparametric statistical procedures for the changepoint problem. *Journal of Statistical Planning and Inference*, 9, 389–396.

[27] Elkie P., Rempel R., & Carr A. (1999) Patch Analyst User's Manual. Ontario Ministry of Natural Resources. *Northwest Science and Technology, TM002*. Thunder Bay, Ontario, Canada, 16.

[28] Frohn R. C. & Hao Y. (2006) Landscape metric performance in analyzing two decades of deforestation in the Amazon Basin of Rondonia, Brazil. *Remote Sensing of Environment*, 100, 237–251.

[29] McGarigal K. & Marks B. J. (1995) FRAGSTATS: spatial pattern analysis program for quantifying landscape structure. *General Technical Report. PNW-GTR-351*. Portland, OR: U.S. Department of Agriculture, Forest Service, Pacific Northwest Research Station. 122 pp.

[30] Juahir H. (2009) Water quality data analysis and modeling of the Langat River basin. Ph.D. thesis, Faculty of Science, University of Malaya, Kuala Lumpur. 68–138.

[31] Hezri A. A. & Hasan M. N. (2004) Management framework for sustainable development indicators in the State of Selangor, Malaysia. *Ecological Indicators*, 4, 287–304.

[32] Wang K., Wang H. J., Shi X. Z., Weindorf D. C., Yu D. S., Liang Y., et al. (2009) Landscape analysis of dynamic soil erosion in Subtropical China: a case study in Xingguo County, Jiangxi Province. *Soil and Tillage Research*, 105, 313–321.

[33] Puncak Niaga Sdn. Bhd. (2008) A report on dam operations and management. Selangor, Malaysia, published online at www.puncakniaga.com.my.

[34] Shaaban A. J. & Low K. S. (2003) Droughts in Malaysia: a look at its characteristics, impacts, related policies and management strategies. *Water and Drainage 2003 Conference*, Malaysia, published online at nahrim.academia.edu

[35] Yenilmez, F., Keskin, F., & Aksoy, A. (2011). Water quality trend analysis in Eymir Lake, Ankara. *Physics and Chemistry of the Earth, Parts A/B/C*, 36(5), 135–140.

[36] Wilson D., Hisdal H., & Lawrence D. (2010) Has streamflow changed in the Nordic countries? Recent trends and comparisons to hydrological projections. *Journal of Hydrology*, 394, 334–346.

[37] Yang Y. & Tian F. (2009) Abrupt change of runoff and its major driving factors in Haihe River Catchment, China. *Journal of Hydrology*, 374, 373–383.

[38] Sahoo D. & Smith P. K. (2009) Hydroclimatic trend detection in a rapidly urbanizing semi-arid and coastal river basin. *Journal of Hydrology*, 367, 217–227.

[39] Khaliq M. N., Ouarda T. B. M. J., Gachon P., Sushama L., & St-Hilaire A. (2009) Identification of hydrological trends in the presence of serial and cross correlations: a review of selected methods and their application to annual flow regimes of Canadian rivers. *Journal of Hydrology*, 368, 117–130.

[40] Hamed K. H. (2009) Enhancing the effectiveness of pre-whitening in trend analysis of hydrologic data. *Journal of Hydrology*, 368, 143–155.

[41] Chang H. (2008) Spatial analysis of water quality trends in the Han River Basin, South Korea. *Water Research*, 42, 3285–3304.

[42] Hamed K. H. (2008) Trend detection in hydrologic data: the Mann-Kendall trend test under the scaling hypothesis. *Journal of Hydrology*, 349, 350–363.

[43] Kahya E. & Kalayci S. (2004) Trend analysis of streamflow in Turkey. *Journal of Hydrology*, 289, 128–144.

[44] Xu Z. X., Takeuchi K., & Ishidaira H. (2003) Monotonic trend and step changes in Japanese precipitation. *Journal of Hydrology*, 279, 144–150.

[45] Hamed K. H. & Ramachandra Rao A. (1998) A modified Mann-Kendall trend test for autocorrelated data. *Journal of Hydrology*, 204, 182–196.

[46] Sundell-Turner N. M. & Rodewald A. D. (2008) A comparison of landscape metrics for conservation planning. *Landscape and Urban Planning*, 86, 219–225.

[47] DiBari J. N. (2007) Evaluation of five landscape-level metrics for measuring the effects of urbanization on landscape structure: the case of Tucson, Arizona, USA. *Landscape and Urban Planning*, 79, 308–313.

Section 3

The Anthropogenic Impacts on Landscape Creation

Chapter 5

The Anthropic Pressure on the Landscapes of the Subcarpathian and Piedmont Basin of Dâmboviţa River (Romania)

Mihaela Sencovici and Gica Pehoiu

Additional information is available at the end of the chapter

http://dx.doi.org/10.5772/63722

Abstract

The Subcarpathian and Piedmont Basin of Dâmboviţa cover an area of 412 km^2 and largely overlap north-western part of Dâmboviţa County, namely the relief units: Gethian Subcarpathians (the Subcarpahians of Argeş), Subcarpathians of Curvature (the Subcarpathians of Ialomiţa) and Getian Plateau (Piedmont of Cândeşti).

Geographic landscape in the area under analysis is closely correlated to ample, complex activities, and to various effects, depending on concrete local conditions, and all these open a large observation field for geographic research.

In the studied area, in time, in the structure and geographic repartition of the categories of land uses, different changes have intervened, especially related to diverse geomorphological processes, the deforestation of some forest areas and the extension of built areas.

Administratively, the Subcarpathian and Piedmont Basin of Dâmboviţa correspond to the territory of 11 communes. The relief, by its morphographic, morphometric, morphogenetic and morphodynamic features, presenting both favorable and unfavorable aspects, represent one of the factors of geographic environment influencing the characteristics of human habitat, and also conditioning the geographic area occupied by human settlements and their features. By using the lands according to his various interests, man has modified more or less intensely the composition and structure of vegetal cover, which influenced the hydrological regime, present modeling processes, quality of the soil, etc., leading to general changes in the structure of geographic landscape.

Keywords: environment, landscape, anthropic influence, subcarpathian unit, land use, vegetation

1. Introduction

The area under analysis corresponds to the Subcarpathian and Piedmont Basin of Dâmboviţa.

In point of its geographic limits, the delimitation between the Subcarpathian and Piedmont Basin of Dâmboviţa and the neighboring geographic units goes through the following altimetric levels:

In the north, the last alpine peaks of Leaota dominate Subcarpathian hills; the delimitation between the two sectors occurs along the alignment of communes Cetăţeni and Pucheni, the limit going through the following summits: Groapa Oii (950 m), Vârful lui Tică (950 m) and Plaiul Găvanei (1250 m). It neighbors Basin of Argeş in western part, the limit being realized by the hills: La Poşta Veche (716 m), Vârtop (790 m), Malu Corbului (795 m), Perilor (744 m), Istrate (660 m) and Tâmpa (522 m).

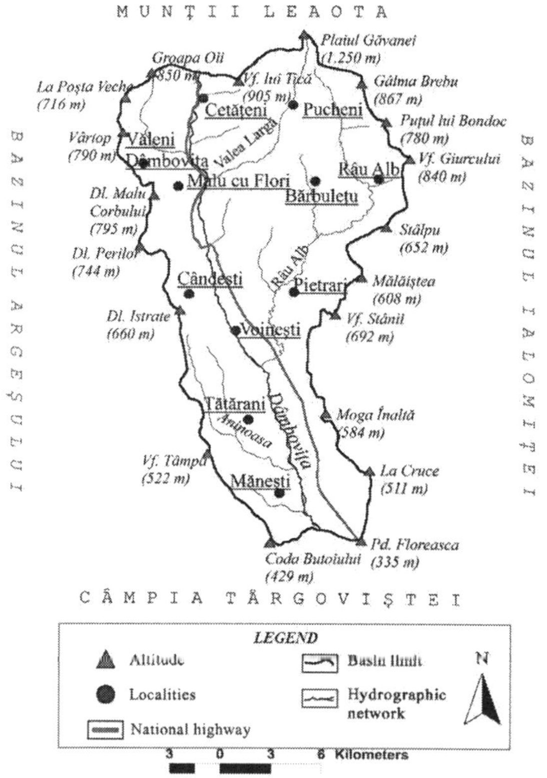

Figure 1. Geographic limits of the Subcarpathian and Piedmont Basin of Dâmboviţa.

http://dx.doi.org/10.5772/63722

In the east, the watershed separating basin of Dâmboviţa from basin of Ialomiţa goes through the following peaks: Gâlma Brebu (867 m), Puţul lui Bondoc (780 m), Vârful Giurcului (840 m), Culmea Stălpu (652 m), Culmea Mălaiştea (608 m), Vârful Stânii (692 m), Culmea Moga Înaltă (584 m) and La Cruce (511 m). In the south, from Dragomireşti, where Dâmboviţa enters the plain, the hill margin goes around the top of a terrace (t3), which vanishes away near the locality of Priseaca. The limit is given by the peak Coada Butoiului (419 m) and the forest Floreasca (355 m)—**Figure 1**.

As topography and morphohydrography, the basin evolves into an elongated form from north to south, being larger in the north, where it receives two important tributaries, namely Râu Alb and Valea Largă, while southwards it is narrower, being part of the plain sector (at Dragomireşti). The hilly area of Dâmboviţa is a region where fruit trees are grown. The Fruit Tree Station of Voineşti, created in 1950, polarized the whole valley beginning from Cetăţeni up to Dragomireşti, the area being crossed by just one national road, DN 72 A, connecting Târgovişte Town to Câmpulung.

The relief represents one of the factors of the geographic environment influencing the features of human habitat [1]. By its morphographic, morphometric, morphogenetic and morphodynamic features, highlighting both favorable and restrictive aspects, the relief conditions the geographic area occupied by human settlements and their features. The hilly basin of Dâmboviţa covers an area of 412 km^2 and is represented both by Subcarpathian Hills and Piedmont Hills. This situation is due to the fact that while on the left side Dâmboviţa passes from the alpine zone to the plain area by means of Subcarpathians, on the right side, in-between Subcarpathians and the plain, lies the Getic Piedmont.

2. Methodology

In the present analysis, we have taken into account the value of human pressure by the way lands are used in agriculture, which represents a synthetic indicator allowing to appreciate the intensity of the impact of anthropic activity on the environment, bringing to light the degree of artificialization of vegetal cover in the area under analysis. Although it is an indicator that is frequently used, its value being quite relative, because pressure is differentiated also depending on the inhabitants' occupations and on the type of agriculture practiced (intensive or subsistence). The formula applied by FAO for the calculation of this indicator is:

$$P = \frac{S(\text{ha})}{N(\text{inh.})}$$

where: P = human pressure; S = area under analysis; N = number of inhabitants in the area under analysis.

Using this formula, we have calculated the human pressure on the environment using various agricultural land uses, namely: arable, pastures, hayfields, vineyards, orchards, for the year

2014, based on data provided by the Romanian National Statistics Institute. Statistical data used in this analysis are on the level of each commune and it is hard to differentiate them for the communes which have only a part of their territory in the area analyzed.

3. Human pressure on the natural landscapes

In time, relation between man and the environment has changed deeply both in the area of Subcarpathian and Piedmont Basin of Dâmbovița and nationally [2].

The influence of human activity on the environment is particularly complex and has various effects. From simple anthropic activities: plant cultivation, animal breeding, wood exploitation, up to the complex ones: oil exploitation, mining, extraction of building materials, along with tourism, arrangement of the infrastructure and increase of constructible fund, all these bring changes concerning natural landscapes [3].

Man, by the totality of his activities, has intervened on the environment even since the oldest times, until now, triggering major changes in the landscape [4].

Demographic growth has led to a significantly increased consumption of natural resources, while economic, social and technical development has led to the appearance of more and more efficient means and techniques of exploration, exploitation and transformation of the raw matters. The last decennia, on the background of a growing consumption, have been characterized by a high level of energy and raw matters use, and storage of wastes coming from a production meant to face a continually larger and diversified demand.

In the anthropized ecosystems, man has deteriorated the biological processes, ignoring the law of self-regulations in the biosphere. Such situations have affected relation between man and nature, and the aggressiveness that man has expressed by his relation to the environment has gradually turned against himself [5].

3.1. Human pressure on the land used in agriculture

The use of geographic area depends on how favorable natural factors are, on the productive potential of land and on improvement works. Agricultural zone represents the terrestrial area exploited by the cultivation of plants, this being an important component of rural area and having certain limits imposed by relief and pedoclimatic conditions.

At the present, agricultural area of the Subcarpathian and Piedmont Basin of Dâmbovița is of 20,961 ha and represents 48.70% of total area of the basin of 43,045 ha (**Figure 2**). On the level of administrative units, the situation is as follows:

By diverse anthropic activities undertaken in the agricultural area, man has brought structural changes in the natural vegetation. From deforestations undertaken to increase the agricultural area up to changes in the makeup of the vegetal cover, through the use of lands for different agricultural cultures, natural landscapes have been submitted to a continual anthropic pressure. The value of anthropic pressure, following the use of lands for agricultural activities, represents an indicator allowing a concise appreciation of the impact of human activity on the environment, highlighting in this way the artificialization degree of vegetal cover in the analyzed area [6].

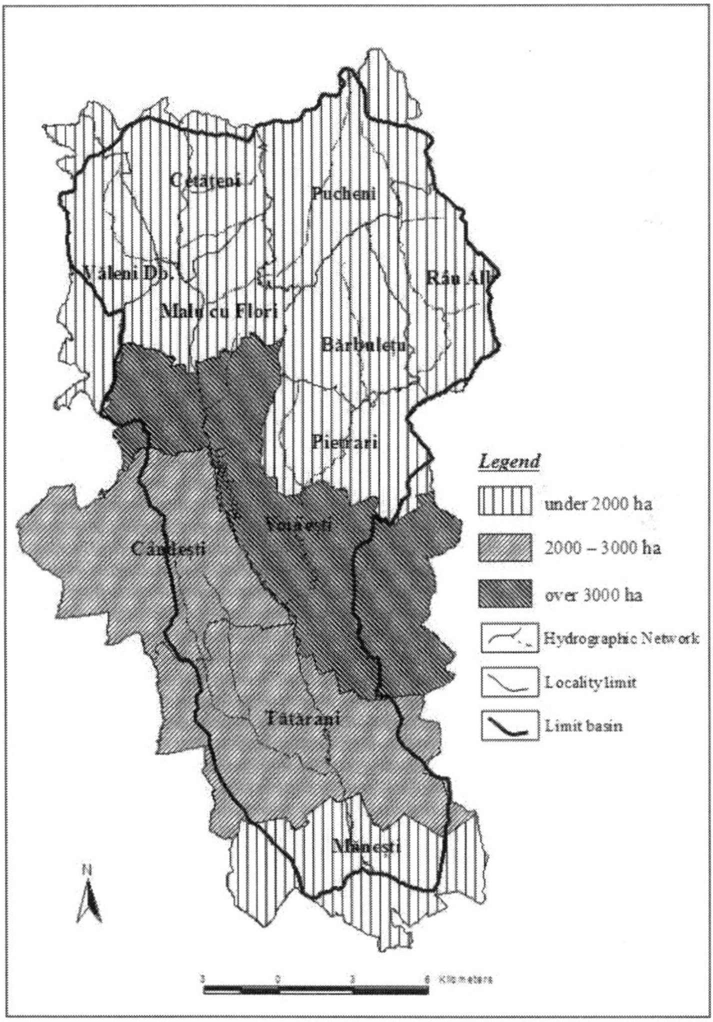

a

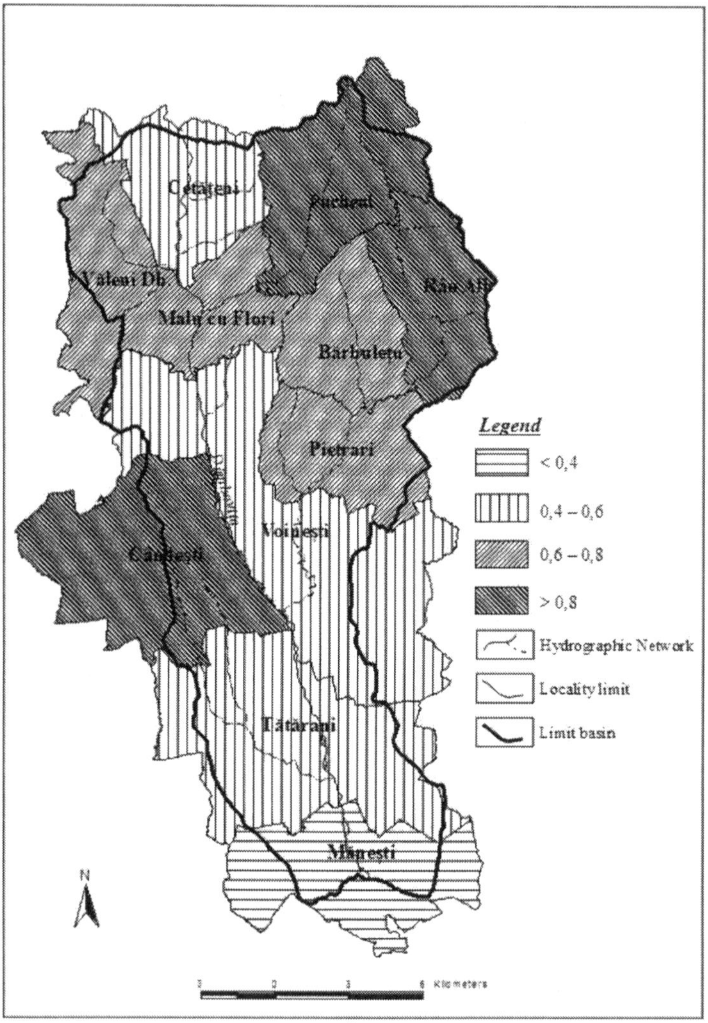

b

Figure 2. (a, b) Agricultural area and human pressure through the agricultural area on the level of administrative units.

The value of this indicator is quite relative, because pressure is differentiated depending on the type of agriculture practiced (intensive or subsistence) [7]. Using the formula applied by FAO for the calculation of this indicator, we have calculated human pressure on the environment through agricultural use of the lands, for the year 2013 based on data provided by National Statistics Institute.

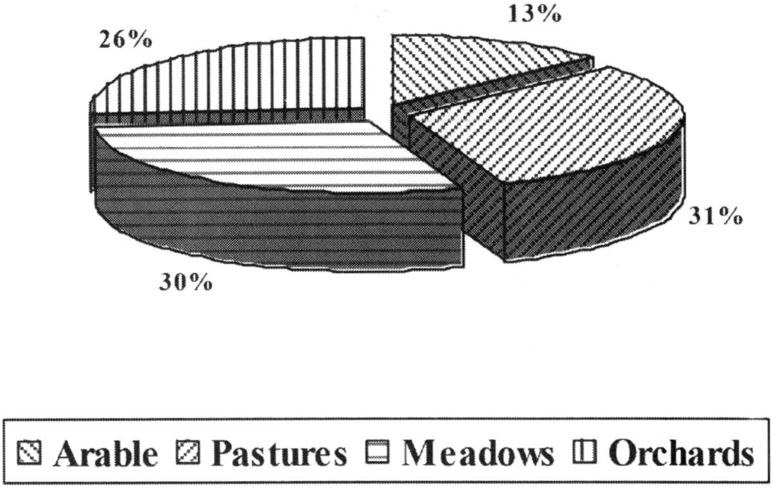

Figure 3. Structure of the agricultural fund—percentage.

The human pressure on the environment was calculated taking into account agricultural area of the zone under analysis divided into administrative-territorial units of the third degree—communes. On the whole, on agricultural landscape in the Subcarpathian and Piedmont Basin of Dâmboviţa, a pressure of 0.68 is exerted. On the administrative level, the highest values of human pressure index correspond to the localities: Râu Alb (1.13 ha/inh.), Pucheni (0.97 ha/inh.) and Cândeşti (0.81 ha/inh.). At the opposite pole, with lowest values of the anthropic pressure, we find the localities: Măneşti (0.24 ha/inh.), Tătărani (0.41 ha/inh.)—**Figure 2**.

In the agricultural area of Subcarpathian and Piedmont Basin of Dâmboviţa, largest sectors are those corresponding to pastures, hayfields and orchards, and to a lower extent, arable zones (**Figure 3**).

3.2. Human pressure on the environment by means of pastures and hayfields

Area corresponding to the zone of pastures and hayfields is 12,825 ha and represents 61.18% of arable area in the zone under analysis, namely 29.79% of the total area. Administratively, largest areas meant for pastures and hayfields are in the localities Cândeşti (1667 ha), Pucheni (1522 ha), and smallest areas belong to the localities Măneşti (351 ha) and Malu cu Flori (678 ha).

The highest values of the index of human pressure on the environment by means of pastures and hayfields correspond to the localities: Râu Alb (0.88 ha/inh.), Pucheni (0.81 ha/inh.), and the lowest values to Măneşti (0.06 ha/inh.), Tătărani (0.22 ha/inh.) while the average pressure on the Subcarpathian and Piedmont Basin of Dâmboviţa is of 0.36 (**Figure 4**).

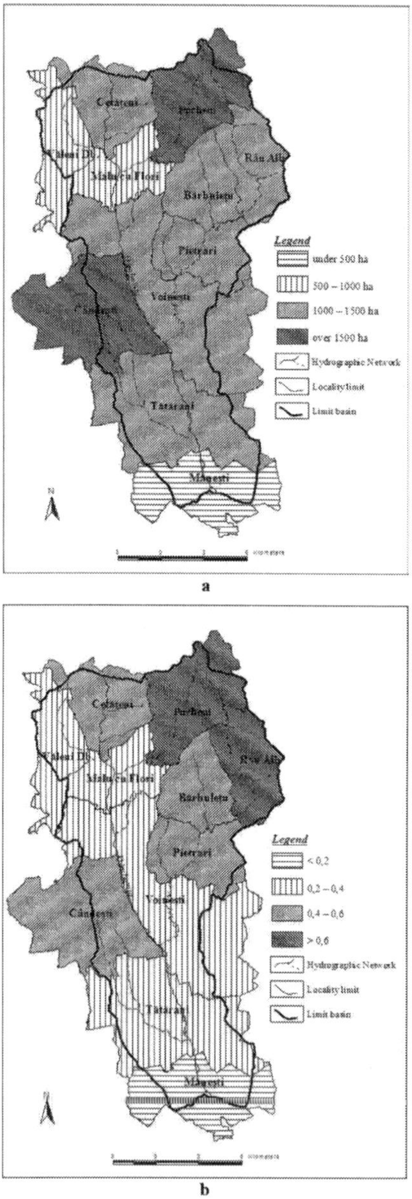

Figure 4. Area of pastures and hayfields (a) and human pressure exerted through them on the level of administrative units (b).

Concretely, areas meant for pastures have undergone pressures of an excessive grazing that triggered the continual degradation of the quality of vegetal cover. A part of the areas corresponding to pastures overlap riverside valley of Dâmboviţa River. Here, following the excessive grazing, a large part of the lawns of *Fetusca valesianca* have turned into lawns of *Poa bulbosa* and *Cynodon dactylon* (**Photo 1**).

For hayfield area, the main type of anthropic pressure is using of chemical fertilizers to increase the quality of the herbaceous cover; consequently, a part of natural vegetal formations and a part of flora were strongly affected.

Photo 1. Grazing in the river meadow of Dâmboviţa (Măneşti), excessive grazing (Gheboieni).

Photo 2. Excessive grazing (Râu Alb).

The degradation of vegetal cover appears in the moment when area of the respective land is overused through daily grazing; such areas can be met in the communes Râu Alb, Pucheni, Cetăţeni and also in some parts of the riverside of Dâmboviţa and of its tributaries (**Photo 2**).

3.3. Human pressure on the environment by means of orchards

Presence of the Fruit Tree Station Voineşti has supported whole zone by means of a strong development of fruit-tree growing sector. So, after the areas occupied by pastures and hayfields, those occupied by fruit trees hold a considerable percentage, more exactly 25.88% of the agricultural area, i.e. 5340 ha. Largest fruit tree areas belong to the localities: Voineşti (1300 ha), Văleni Dâmboviţa (840 ha), Malu cu Flori (769 ha), and smallest areas can be found in the localities Tătărani (162 ha) and Cetăţeni (173 ha)—**Figure 5**.

The human pressure exerted by areas occupied by orchards on whole area of the basin is of 0.15 ha/inh. On the administrative level, localities with a high index of the anthropic pressure are: Malu cu Flori (0.33 ha/inh.), Văleni Dâmboviţa (0.31 ha/inh.), the lowest index is that of the localities Tătărani (0.03 ha/inh.), Cetăţeni (0.05 ha/inh.) and Măneşti (0.07 ha/inh.)—**Figure 5**.

During last few years, by accessing European funds, a part of the traditional and of intensive fruit-tree areas have been replaced by superintensive ones (**Photo 3**).

Photo 3. Superintensive orchards with drop by drop irrigation systems (Voineşti).

Having a superior production level, their maintenance is also different, superintensive fruit-tree areas exerting a much higher pressure on the environment, compared to traditional ones. The main form of pressure on the natural environment is represented by use of large quantities of chemicals, especially in areas occupied by apple trees. Approximately 70% of all orchards correspond to areas occupied by apple trees and pear trees, about 4000 ha. For just

1 ha of apple trees are used, on average: 20 g insecticide, 40 g fungicide and 1000 l water for just one hygienization treatment. Consequently, if, on average, depending on the meteorological conditions, 15 treatments/year are needed, referring to the area of 1 ha, we can notice a consumption of water of 15,000 l. For just one agricultural year, fruit tree area (apple and pear trees) of the Subcarpathian and Piedmont Basin of Dâmboviţa requires 60,000,000 l water (60,000 tons of water).

Because of the lack of collaboration between orchard owners, there has occurred a situation in which hygienization treatments came to be repeated even every 4 days, whereas the period recommended is 7 days. Quite often, orchards had to suffer following intoxications with chemicals.

Differences between the superintensive and traditional and even intensive plantations, regarding to the degradation of vegetal cover, are significant. In superintensive plantations, with drop by drop irrigation systems, vegetal cover is almost totally missing, because of plowing works, combined with the use of herbicides (Photo 3).

At the actual rhythm, a part of fruit tree area of the localities situated in the riverside of Dâmboviţa (Voineşti, Cândeşti, Malu cu Flori) shall be occupied by superintensive orchards, changing well-known landscape offered by traditional and intensive orchards. The localities with more significant fruit tree areas, yet having no flat land, but situated on slopes, will preserve their aspect, at least in short and medium term, due to difficult installation of a drop by drop irrigation system (**Photo 4**).

Photo 4. Traditional apple orchard: 1. Pietrari, 2. Plum orchard (Bărbuleţu).

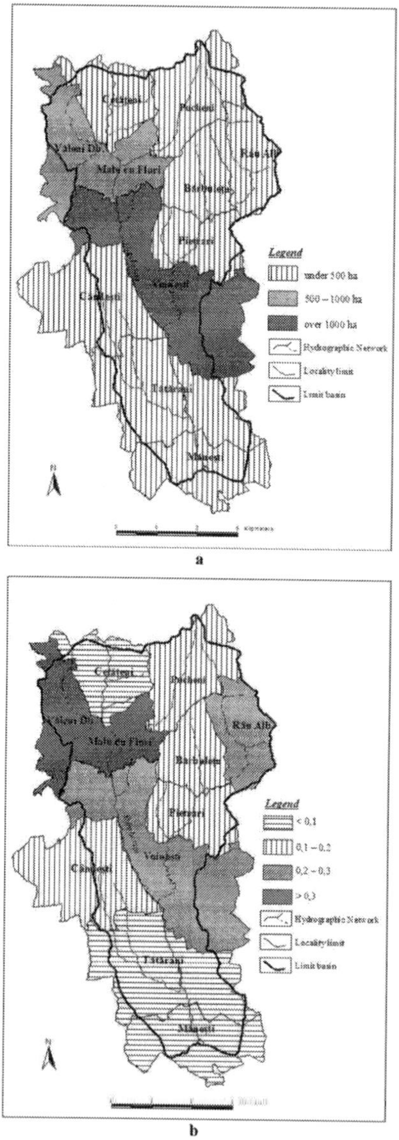

Figure 5. Area occupied by orchards (a) and human pressure (b) exerted by means of orchards on the level of administrative units.

http://dx.doi.org/10.5772/63722

3.4. Human pressure on the forest areas

Out of all land ecosystems, forests represent a special category, given the raw matters they provide, natural possibilities of regeneration and multiple services having a protective character. Forest represents a favorable environment of life for different species of plants and animals, having a significant hygienic-sanitary role, exerting important beneficial influences on climate and soil, and it is an important tourist element.

In fact, it influences the climatic regime through the improvement of most climate factors, such as temperature, radiations, precipitations, evaporation, air humidity, evapotranspiration, wind, turning the forest into a unique environment. Forest has an important role in cleaning, purification of the atmosphere, by means of photosynthesis process, by massiveness of its rich foliage, having the characteristic feature of freeing a large quantity of oxygen, absorbing carbon dioxide, and retaining vapors and toxic gases, dusts and sound waves [8].

All these functions are part of the category of protective functions of the environment (soil, waters, climate, and ambient atmosphere), having rather indirect influences on the human society. Forest also has numerous direct influences, by its social protection functions: recreational, esthetic and landscape function or as an object of study due to the large quantity of scientific information it can provide. It should be noted that forest is the ecosystem assuring most complex and stable protective balance for natural environment. It realizes ideal connections between flora, soil and climate, developing tall and long-lasting trees with an important ecological function which determines a network of compensations and self-regulations in the biosphere.

Yet, of all terrestrial ecosystems, most deteriorated have been forest ecosystems, following destructive actions undertaken by man, both by turning to good use the raw matters and by replacing them with other, less enduring, ecosystems. A part of the forest area has been deforested to enlarge agricultural and constructible areas, and on the other hand, another part has been degraded through abusive exploitations, fires, permanent grazing, pollution [5].

At the present, forest area of the Subcarpathian and Piedmont zone of Dâmbovița covers 19,436 ha and represents 45.15% of the area of this basin. It is made up of numerous deciduous species: *Fagus silvatica, Carpinus betulus, Fraxinus excelsior, Sorbus terminalis, Quercus patraea, Acer platanoides, Acer capestris, Ulmus procera, Tilia platyphylos, Quercus robur, Quercus cerris* which cover Subcarpathian Hills and the Piedmont, along with *Alunus glutinosa, Populus alba, Populus nigra, Salix fragilis* developed in the river plains of Dâmbovița and its tributaries.

From an administrative perspective, largest forest areas are in the communes: Voineşti (4362 ha), Tătărani (3926 ha), Cândeşti (2735 ha), and the smallest in the communes Malu cu Flori (415 ha), Văleni Dâmbovița (683 ha) and Râu Alb (723 ha)—**Figure 6**.

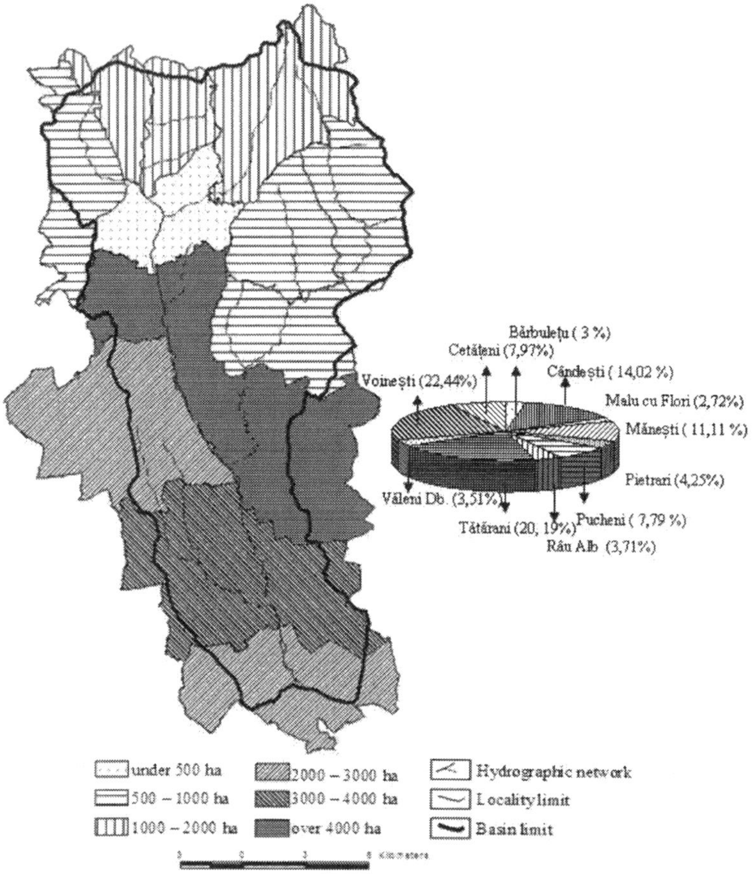

Figure 6. Forest area.

Based on these data, one can appreciate that in the area under analysis, namely the zone of Subcarpathian and Piedmont Basin of Dâmbovița, forests develop on a significant area, about half of the total area (45.15%); therefore ecological balance is maintained. Yet, anthropic intervention on the forest zones, even though on relatively small areas, is continuous. Main types of pressure, of deterioration and modification of forested areas are: deforestations, both for wood exploitation and for extension of agricultural zones, mainly by orchards; storage of domestic and agricultural waste, especially at the margin of the forests; fires, excessive grazing leading to destruction of young trees and of vegetal cover.

Out of the need to turn to good value of the wood and to enlarge agricultural and constructed areas, in time, important forest zones have been deforested.

At the present, in the region under analysis, main areas affected by deforestations are those situated at the contact with agricultural zones, especially at the contact with orchards. Here, every year, small areas in the margin of forests are deforested by cutting, by fire or by use of herbicides to increase areas meant for fruit trees. Such practices are encountered in almost entire Subcarpathian and Piedmont Basin of Dâmboviţa, especially in areas of the communes Malu cu Flori, Pietrari, Cândeşti, Bărbuleţu and Voineşti (**Photo 5**).

Photo 5. Deforestations used to extend the areas meant for hayfields and orchards.

Another category of deforestations is that in which wood is exploited to be turned to good value; unlike the deforestations intended for increasing of agricultural area, they take place on larger areas and in the forest zones.

Out of desire to increase fruit tree areas, especially those of apple trees, there have been regions deforested on slopes; this, together with excessive humidity and pedological conditions, has triggered landslides [3]. Such sectors, where geomorphological processes and landslides are more marked, affecting environment and anthropic activities, are present in the localities Bărbuleţu, Pietrari, Râu Alb, and Pucheni (**Photo 6**).

Photo 6. Landslides triggered by the replacement of the forest areas by orchards (Pietrari).

Area of Pietrari Commune situated at the border with Bărbuleţu Commune presents precisely consequences of uncontrolled deforestations and of attempt to replace forest vegetation by fruit trees. Pedological conditions, abundant rains, and incapacity of the fruit trees to retain humidity and to stabilize the soil—compared to forests—have triggered landslides. Such situations can be met as well in the communes Râu Alb, Bărbuleţu, Văleni Dâmboviţa, and Pucheni (**Photo 7**).

Photo 7. Landslides triggered by uncontrolled deforestations.

http://dx.doi.org/10.5772/63722

Human pressure by waste storage in forest areas. This phenomenon, namely depositing domestic and agricultural wastes in the forest areas does not affect large zones, due to domestic waste collection program [9]. Areas affected by waste storage can be met at the margin of forests and at the contact of latter with a river. Due to the fact that a significant part of domestic wastes are hardly biodegradable, without an intervention meant for cleaning the respective zones, they will continue to affect environment and natural landscape (**Photo 8**).

Photo 8. Waste deposited at the margin of the forests.

Another pressure phenomenon on the forest areas is grazing. Increasingly larger areas of the forest zone are affected by this phenomenon, destroying cover of small vegetation, both by breaking young trees and shrubs and by treading on vegetal cover (**Photo 9**).

Photo 9. Grazing in forest areas.

3.5. Human pressure on the hydrographic network

Regarding human pressure on the hydrographic arteries in Subcarpathian and Piedmont Basin of Dâmboviţa, we have identified two major problems: pollution and increased water consumption following agricultural activities.

Photo 10. Domestic waste stored near the rivers.

Photo 11. Storing the packaging of chemicals waste in the riverside.

http://dx.doi.org/10.5772/63722

Water pollution represents the alteration of its physic-chemical and biological properties, following direct or indirect human activities, so that water becomes unusable for consumption. Main pollution sources for the hydrographic arteries are depositing on the riverside of domestic and agricultural wastes (**Photo 10**). To these, one can add packaging of chemicals (insecticides, pesticides, fungicides, herbicides) used in the hygienization treatments of plantations; consequently, flora but also fauna are strongly affected (**Photo 11**).

Along the whole riverside of Dâmbovița and its tributaries situated near orchards and village precincts are stored packages with toxic substances and domestic wastes, affecting the environment and numerous organisms of water environment. The wrong storage also triggers possibility of blocking watercourse, when hydrological regime is increased; consequently, there is a risk of flooding some agricultural areas, houses and roads.

Photo 12. Blockage of the course of Râu Alb rivulet following the storage of domestic wastes on its riverside.

Such a situation was recorded in the year 2012, when following abundant rains, the rivulet Râu Alb increased its flow, and at the passage through Valea Village (Pietrari Commune), because of domestic and agricultural wastes deposited on the wrong way, normal course of water was blocked, flooding agricultural areas, the communal road of Valea Village (Pietrari) and households on the opposite side of protective dam (**Photo 12**). The increasing water quantity needed for agricultural activities, especially for fruit-tree growing, together with high temperatures of the interval June–August, have led to appearance of drying phenomenon on numerous permanent hydrographic arteries, in certain sectors. Entire area meant for fruit trees —apple trees and pear trees—uses on average about 60,000 tons water/year only for hygienization and maintenance, to which one can add water used in drop by drop irrigation systems, from increasingly numerous superintensive plantations. Excessive grazing in the riverside of Dâmboviţa and of its tributaries constitutes another form of pressure on vegetal cover and young or small trees. This, along with treading of water course margins by animals and agricultural equipments, determines a decreased resistance to water erosion. Riverside of Râu Alb, in between villages Valea Câmpului (Pietrari) and Manga (Voineşti), represents an area in which pressure of excessive grazing, combined with abusive creation of a road used by heavy agricultural equipments, have led to the deterioration and even complete elimination of small trees and of vegetal cover that used to support the right riverside.

4. Conclusions

By using lands according to his needs and interests, man has triggered the dwindling of areas occupied by natural vegetal formations and their replacement by agricultural cultures, secondary vegetal formations or lands that have become unproductive through degradation. On the other hand, anthropic activity has triggered the modification of structure of vegetal cover where natural vegetal formations were maintained. All these have influenced hydrological regime, present relief modeling processes, quality of soil, leading to general modifications in the structure of geographic landscape.

Natural ecosystems are the ones that have been changed little constitute richest resource in point of organisms, biocenoses, environmental conditions, and also numerous relations between organisms and environmental factors, food chain and networks. In the case of natural ecosystems, between the populations of a biocenosis there appear some self-regulation mechanisms, and so organisms that no longer find their ecological niche are obliged to emigrate or disappear.

Super-intensive fruit tree areas exert a much higher pressure on environment, compared to traditional fruit tree areas, through use of large quantities of chemicals, especially in apple tree areas. For just 1 ha of apple trees are used, on average, 20 g insecticide, 40 g fungicide and 1000 l water for a single hygienization treatment.

Using of chemicals affects both flora and fauna of the respective area. A more serious collaboration between fruit growers, and respect for warnings of the specialists based on meteorological conditions could favor diminution of the quantity of chemicals used and decrease of annual water consumption needed to hygienize plantations.

Main types of pressure, impact on and modification of forested areas come from: deforestations, both for using wood and for extending agricultural areas, mainly orchards; domestic and agricultural wastes storage, especially at the margin of forests; fires, excessive grazing, leading to destruction of young trees and of vegetal cover.

In the area under analysis, zones affected by deforestations are those situated at the contact with agricultural areas, especially at the one with orchards (through cutting, fires or use of herbicides to enlarge areas meant for fruit trees).

Acknowledgements

This material is only the contribution of its authors and it was not published anywhere else.

Author details

Mihaela Sencovici and Gica Pehoiu*

*Address all correspondence to: gpehoiu@yahoo.com

Valahia University of Târgovişte, Faculty of Humanities, Târgovişte, Dâmboviţa County, Romania

References

[1] Bălteanu, D. Semnificaţia geografică a modificării utilizării terenurilor (Geographic significance of land use changes). Bucureşti: AUŞM-Geol.-Geogr. 1996; V: 5-9.

[2] Muică, C. Influenţa modului de utilizare a trenului asupra dinamicii peisajului (Land use influence on landscape dynamics). Bucureşti: Terra. 1991; 2-4.

[3] Muică, C., Sencovici, M. Floristic Diversity and Human Pressure in the Subcarpathians. Bucureşti: Analele Univ. Hyperion (Annals of Hyperion University). 2006; tome IV–V: 57–64.

[4] Bertrand, G. Paysage et la geographie physique globale. Esquisse methodologique. Toulouse: Revue Geographique des Pyrenees et du sud-oust. 1968; 39, 3.

[5] Sencovici, M. Studiul geografic al mediului în câmpia Târgoviştei (A Geographic Study of the Environment in The Plain of Târgovişte). Târgovişte: Editura Transversal; 2010: ISBN 978-606-8042-59-6.

[6] Muică, C., Zăvoianu I. Relationship between the state of plant cover, runoff and erosion processes, in "Vegetation. Land Use and Erosion Processes", Bucureşti: Institutul de Geografie (Institute of Geography). 1999; 88-94.

[7] Dumitraşcu, M. Indici utilizaţi în evaluarea gradului de transformare antropică a peisajelor din Câmpia Olteniei (Indexes Used in the Evaluation of the Anthropic

Transformation Degree of the Landscapes of the Plain of Oltenia). Bucureşti: Revista Geografică (Geographic Review). 2005; XI: 68-73.

[8] Pehoiu, G., Muică, C., Sencovici, M. Geografia mediului cu elemente de ecologie, Târgovişte: Editura Transversal; 2006: ISBN: 978-973-7798-32-9.

[9] Sencovici, M. Vegetal communities in Dâmboviţa County. The Annals of University of Târgovişte; 2009; 10: 43-50.

Chapter 6

Tracking Anthropogenic Influences on the Condition of Plant Communities at Sites and Landscape Scales

Richard Thackway

Additional information is available at the end of the chapter

http://dx.doi.org/10.5772/62874

Abstract

Deriving vegetation condition assessments from land use classifications and mapping only provides rudimentary and very coarse insights; ones that may be misleading for planning and priority setting by regional planners and policy makers. Standardized indicators of past ecological resilience for a particular landscape can assist land managers and ecologists track, evaluate and report the outcomes of land management decisions. Developing a timetable of the varying goals of land management practices and their past effects on vegetation condition including regenerative capacity is suited to on-ground managers.Access to information about how and why landscapes were transformed helps regional planners and policy makers identify and prioritise areas for investment relative to an ideal state. Tracking the responses of previous native plant communities to a range of land management practices helps decision makers gain an understanding of which outcomes can be realistically achieved in particular landscape contexts.

Keywords: Tracking change, Plant communities, Vegetation condition, Management effects, Transformation, Resilience

1. Introduction

Anthropogenic management of vegetated landscapes, commonly linked to the delivery of food and fibre, normally changes components of vegetation condition; structure, composition and function of plant communities [1]. Land use and land management practices manipulate ecological patterns and process resulting in various, and sometimes drastic changes to the components of vegetation condition patterns, both spatially and temporally. Repeated interventions usually deliberate but sometimes inadvertent, transform landscapes over time.

As a result, used by anecologist have been modified and fragmented to varying extents by modern human use and management.

Land management practices, and thus the influences these practices place on criteria and indicators of vegetation condition at site and landscape scales, change over time. Therefore, the responses of landscapes and the landscapes themselves also change over time. Past land use decisions strongly influence current condition, which in turn strongly influence opportunities for future use and management [2].

Depending on the landscape context and the landscape's genesis, land use and land management practices can change or manipulate ecological patterns and process resulting in various, and sometimes obvious changes in land cover patterns, both spatially and temporally. Repeated interventions, involving both deliberate and/or inadvertent management changes to the components of vegetation, can transform landscapes over time. Other anthropogenic factors can influence the health and vitality of plant communities, including the periodicity and severity of stress and disturbance regimes, and the sequences of different impacts (natural and human initiated). In some cases, the management of native vegetation involves an almost complete removal of the components of vegetation condition; structure, composition and function, for example irrigated crops. Other landscapes are minimally managed resulting in no obvious effects on the components of vegetation condition over time, for example conservation and protected areas.

Land planners and managers require information on the status, change and trend of these impacts on vegetation for environmental reporting, land use trade-offs, and to inform future land use scenarios. On-ground managers need more temporally orientated detailed information for the area of ground under their control to inform their operational planning, management and monitoring. While regional and state-wide land use policy and planners need more spatially than temporally orientated landscape scale information to inform decisions on priority setting, setting targets and selecting areas and sites for monitoring, evaluation and reporting. Fundamentally, it is the same space-time-point information but the information needed which has to be presented in a different context and be fit for the intended purpose.

The time to start documenting land use and land management histories is critical. The point at which human action becomes a major factor in environmental change varies with time, location, and culture. Pre- and post-European settlement in many countries represents a key point from which to assess the transformation of vegetated landscapes including changes in vegetation condition. For convenience and to align with various international agreements, in Australia, this date is usually defined as 1750, around the time of the industrial revolution [3]. In Australia, this date is prior to European settlement, in 1788, and thus reflects the pre-European land-use status, which although not without substantial human effects on the landscape, had been relatively stable for many hundreds of thousands of years [4]. Over the last decade, considerable progress has been made in developing information systems for monitoring and reporting the current extent and condition of vegetation types, relative to a pre-European unmodified state [5–8].

http://dx.doi.org/10.5772/62874

Native vegetation is commonly used as the key descriptor of landscape type, extent and condition. Indicators of vegetation structure, composition and regenerative capacity are widely used to monitor and report sustainable production and biodiversity conservation [9, 10]. However, there is a little agreement on standardised methods for how best to track and report anthropogenic influences on the condition of plant communities at sites and landscape scales, including how land use and land management practices have been used to transform, and continue to transform, the extent and condition of vegetated landscapes. The absence of a consistent approach to monitor and report changes in native vegetation condition over space and time remains a source of contention, and even conflict, between those involved in conservation and protection and those involved in sustainable land use and management of native vegetation [11].

Land use purposes include wood production, biodiversity conservation, water quantity and quality, agriculture production (cropping) and maintaining sustainable grazing systems [12]. Management practices associated with each land use are used to change or modify the ecological building blocks by: modifying, removing and replacing, enhancing, restoring, maintaining and/or improving components of vegetation condition; structure, composition, function [13]. The intent of use and management of vegetated landscapes is commonly linked to the delivery of food and fibre and is affected by changing or manipulating the components of vegetation condition, structure, composition and function of plant communities [1]. As a result, most landscapes have been modified and fragmented to varying extents by modern human use and management.

Integrated spatial and temporal monitoring and reporting of the effects of anthropogenic interventions using standardised assessment and tracking tools can offer new insights into the effects that past socio-economic drivers and their effects on the current condition of sites and landscapes, as well as predicting the likely future condition states expected as a result of changing the goals of management actions and their intended or likely effects on criteria and indicators of vegetation condition; structure, composition and function of plant communities. Such standardised indicators of resilience assist land managers and ecologists track, evaluate and report the outcomes of land management interventions.

This chapter highlights the value of documenting spatial and temporal changes caused by anthropogenic changes to the condition of native vegetation. Access to the long-term data and information on changes and trends in vegetation condition is discussed. Two broad approaches are outlined for generating information on vegetation condition; (1) land use classification and maps and iand management regimes suitable for the regional planner and policy maker and (2) developing detailed site-based chronologies of land management practices and their observed effects on criteria and indicators of vegetation condition more suited to on-ground managers. The benefits and shortcomings of both approaches are discussed.

2. Sources of vegetation condition information

2.1. Land use and land management regimes

Six broad land use classes have been defined for surveying, classifying and mapping land-related land uses in Australia [14]. Classes 1 and 2 relate to the use of native vegetation. Class 1 defines those areas where the primary purpose is minimal use; this includes areas protected primarily for nature conservation. Class 2 defines those areas where the primary purpose includes the sustainable use and management of native vegetation for food and fibre production. Classes 3 and 4 define those land use types where the primary purposes are intensive agricultural production, both dryland and irrigated cropping and forest plantations. Class 5 defines the built environment and infrastructure. These five classes represent level 1, with more detailed sub-classes defined in levels 2 and 3 [14]. Class 6 is water, which includes natural water bodies and those which are built infrastructure, that is part of the human environment. For the purposes of this chapter, that is monitoring and reporting anthropogenic influences on the condition of terrestrial plant communities, class 6 has not been described.

Land use classes 1–5 only describe the purpose for the use of the land and not how it is managed (refer to case study 1). It is an understanding of how the land has been and is managed that will reveal what effect these management practices have had in maintaining, enhancing, restoring, or removing and replacing native vegetation over time.

Because land use and condition can change at different scales, for example a map or site survey data, it is important to know that when the land use data and information were collected and recollected and at what intervals because land use classifications make no assumptions about the resilience of native vegetation; this is particularly the case in agricultural and landscapes used for forest plantations. Great caution is needed in using land use information to derive insights into native vegetation condition. Nevertheless, the types of land use may be reclassified to provide rudimentary information on the likely condition of native vegetation types, based on the types of land management regimes associated with each land use class and the likely effect that these regimes have had on criteria associated with condition of native vegetation, that is its structure, composition and functional characteristics.

Vegetation condition information derived from land use information, particularly for land use classes 3–5, only represents a 'first approximation' because the land use classification for classes 3–5 assumes that native vegetation is extirpated and the resilience of the components of vegetation has been lost (**Figure 1**) [14]. Even though a landscape has been, or is, used and managed under land use classes 3–5, evidence shows that depending on how the land was managed and for how long, key components of native vegetation condition, that is structure, composition and function may still may be present [13].

Land management practices used in the management of native vegetation under different types of land use reveals that six broad land management regimes (**Table 1**). Land management regimes affect the condition of the native vegetation by maintaining, modifying, changing, enhancing, removing or replacing one or more of the 10 key criteria (**Table 1**). Each land management regime is underpinned by a suite of land management practices, which can

http://dx.doi.org/10.5772/62874

be used by an ecologist to infer whether the broad intent of the land management regime is to maintain, recover, restore, enhance or remove and replace the plant community.

	Land management regimes used in managing native plant communities					
Criteria that are either deliberately or inadvertently changed or modified using land management practices	**No intervention**	**Harvest biomass, seeds, fruit, fibre**	**Encourage regeneration, enhance growth, maturity and reproduction**	**Monitor health and vitality**	**Re-establish, restore and rehabilitate**	**Degrade, remove and/or replace**
1. Soil hydrological status (Fn)	1 and 2	2	1 and 2	1 and 2	1 and 2	2, 3, 4 and 5
2. Soil physical status (Fn)	1 and 2	2	1 and 2	1 and 2	1 and 2	3, 4 and 5
3. Soil chemical status (Fn)	1 and 2	2	1 and 2	1 and 2	1 and 2	2, 3, 4 and 5
4. Soil biological status (Fn)	1 and 2	2	1 and 2	1 and 2	1 and 2	3, 4 and 5
5. Fire regime (Fn)	1 and 2	2	1 and 2	1 and 2	1 and 2	2, 3, 4 and 5
6. Reproductive potential (Fn)	1 and 2	2	1 and 2	1 and 2	1 and 2	2, 3, 4 and 5
7. Overstorey structure (St)	1 and 2	2	1 and 2	1 and 2	1 and 2	2, 3, 4 and 5
8. Understorey structure (St)	1 and 2	2	1 and 2	1 and 2	1 and 2	2, 3, 4 and 5
9. Overstorey composition (Co)	1 and 2	2	1 and 2	1 and 2	1 and 2	2, 3, 4 and 5
10. Understorey composition (Co)	1 and 2	2	1 and 2	1 and 2	1 and 2	2, 3, 4 and 5

Co, composition; Fn, function; St, structure..

Table 1. Groupings of land use classes (i.e. cells) classified by land management regimes and key criteria of vegetation condition that that are either deliberately or inadvertently changed or modified over time.

Numbers 1–5 in the cells of **Table 1** refer to the five Level 1 land use classes described above. These land use classes and the corresponding land management regimes which affect the criteria of native vegetation condition are described below.

2.1.1. *Land use class 1: conservation and natural environments*

Generally management regimes on land gazetted for biodiversity conservation, wilderness areas and urban water catchment areas do not modify or change the ten criteria (**Table 1**) from the reference state. An important exception to this is the effect of deliberately excluding fire, which is a natural disturbance agency in many ecological communities; excluding fire can lead to major changes in native vegetation structure and composition [15].

Land management regimes associated with this land use tend to maintain and/or enhance the structural, compositional and ecological functions close to the unmodified reference state by restricting access, preventing harvesting, minimal control and intervention in managing and or reinstating the natural fire regime, the control and intervention in managing weeds and feral animals, and not intervening with natural recruitment, succession or senescence of the dominant structuring species present or known to have occurred in one or more of the dominant vegetation strata.

The primary land management regimes used in landscapes that are managed for conservation and natural environments include:

- Restoration and rehabilitation
- Enhance growth, maturity and reproduction
- Monitor health and vitality
- No interventions

By inference, landscapes that have consistently retained their native vegetation cover over time have the closest to 'natural' vegetation condition, that is the 10 structural, compositional and ecological functions (**Table 1**) are essentially unmodified.

2.1.2. *Land use class 2: production from relatively natural environments*

Generally land uses classified as managed for defence infantry training areas, state recreation areas near urban settlements, grazing native vegetation, for example rangelands, and timber production in native forests, minimally modify or do not modify the function in the long term (i.e. criteria 1–6 in **Table 1**) and modify to varying extents the structure and/or composition of plant communities (i.e. criteria 7–10 in **Table 1**).

Effects of these land uses are reversible in the short term, that is regenerative processes can be reinstated towards a reference state. Under these land uses, the structural, compositional and ecological functions are variously modified or maintained and enhanced by silvicultural practices at different times, managing or preventing the natural fire regime, controlling selected invasive species (weeds and feral animals) and by manipulating natural recruitment, succession and senescence.

The primary land management regimes used in landscapes that are managed for production from relatively natural environments include as follows:

- Restoration and rehabilitation
- Enhance growth, maturity and reproduction
- Harvest biomass or productivity
- Monitor health and vitality
- No interventions

By inference, land use types that more or less consistently retain the cover of native vegetation over time have higher levels of landscape regenerative capacity, that is even though structural and compositional components may have been modified, the ecological functions have not been altered.

2.1.3. *Land use class 3: production from dryland agriculture and plantations*

Generally, land uses classified as managed for dry land/rainfed cropping, recently planted tree plantations and horticultural systems involve the complete removal of structural and compositional criteria (i.e. 7–10 in **Table 1**) and moderate to major modification of functional criteria (i.e. 1–6 in **Table 1**).

Moderately, intensive land use includes dry land/rainfed cropping, recently planted tree plantations and horticultural systems. These land uses remove and replace the structural and floristic components while some ecological function is maintained or enhanced through the addition of water, nutrients, monoculture species, weedicides, fungicides, insecticides.

Native vegetation is generally degraded, removed, replaced or reduced to remnants in land use classes 3, 4 and 5.

By inference, land use types that remove the cover of native vegetation (i.e. all structural and species indicators) over time have reduced levels of landscape regenerative capacity and the ecological functions have been altered.

2.1.4. *Land use class 4: production from irrigated agriculture and plantations*

Generally, land uses classified as land managed for production from irrigated agriculture and plantations involve the complete removal of structural and compositional criteria (i.e. 7–10 in **Table 1**) and major modification of most functional criteria (i.e. 1–6 in **Table 1**).

Land uses include improved pasture where one or more exotic species of grasses and/or herbs are introduced and managed to achieve greater levels of production than could achieved by managing native pasture species, for example irrigated lucerne pastures and irrigated cropping systemsin south-eastern Australia. Under this land use type, the structural and floristic components are removed and replaced with local non-indigenous structural and or floristic elements while maintaining or augmenting ecological function through the addition of water, soil nutrients, and/or selected overstorey and understorey species.

Native vegetation is generally degraded, removed, replaced or reduced to remnants in land use classes 3, 4 and 5.

By inference, land use types that remove the cover of native vegetation (i.e. all structural and species indicators) over time and modify and replace the nutrient, hydrological and physical status of the soil have very low levels of landscape regenerative capacity.

2.1.5. *Land use class 5: intensive uses (built landscapes)*

Generally, land uses classified as land managed for intensive uses involve the complete removal of structural and compositional criteria (i.e. 7–10 in **Table 1**) and major modification of all functional criteria (i.e. 1–6 in **Table 1**).

Land uses include intensively managedanimal production systems e.g. feedlots, manufacturing and industrial complexes, urban areas and infrastructure, water reservoirs/impoundments, open-cut mining and waste treatment. Under this land use type, the structural, compositional and ecological functions are removed or prevented.

Native vegetation is generally degraded, removed, replaced or reduced to remnants in land use classes 3, 4 and 5.

By inference, land use types that remove the cover of native vegetation (i.e. all structural and species indicators) over time and modify and replace the nutrient, hydrological, biological and physical status of the soil have the lowest levels of landscape regenerative capacity.

In summary, compiling land use classes over time for a defined soil–landscape map unit and in turn reclassifying these land use classes into land management regimes and what criteria are primarily affected, for example vegetation structure, composition and function, can give insights into the mechanisms that are used in maintaining, enhancing, restoring, or removing and replacing native vegetation over time.

This approach to describing and classifying the condition of native vegetation only enables a generalised and very coarse understanding of the drivers for modifying condition and fragmenting native vegetation. By itself, deriving vegetation condition information from land use information falls far short of what information that land planning and manager stakeholders need to know about the changes and trends in the condition of the native vegetation.

A failure to collect and classify which land management practices that are associated with each land use and to compile and collate the effects that these land management regimes have on criteria and indicators of vegetation condition, that is ecological function (Fn), vegetation structure (St), species composition (Co) deems such derived information of little value to a multiple decision-makers.

Where land planners and managers require information on the status, change and trend of these impacts on vegetation types for environmental reporting, land use trade-offs, and to inform future land use scenarios, more detailed information is needed at the soil–landscape level, that is the operational scale of the land manager. Unless such information is collected in association with on-ground land managers, this will deem inferred vegetation condition information derived land use of little or no use to land managers and on-ground decision-makers. For example, to engage on-ground land managers, in the identification and selection of priority areas for restoration and regeneration of a soil-landscape's indigenous plant communities, and to make recommendations on what land management practices might be changed, it is vital to understand what, when and where the land manager/s have changed and/or modified key criteria and indicators of vegetation condition, that is structural, compositional and ecological functions and over what time periods.

An alternative approach that more adequately satisfies these operational requirements of land planners and managers is presented below.

2.2. Developing detailed chronologies of land management practices and their observed effects

Developing a systematic and comprehensive chronology of land management practices and their effects on criteria and indicators of vegetation condition provides a site-based approach to determining anthropogenic effects on the condition of plant communities.

In managed landscapes, vegetation is either deliberately and/or inadvertently modified or fragmented over time by modifying key criteria and indicators including: natural regimes of fire, soil hydrology, soil structure, nutrients and biology and/or the reproductive potential of the native species in overstorey and/or understorey. Where land management regimes and their associated land management practices, at sites and across landscapes, have been prolonged and intensive with the aim or extirpating the native vegetation (**Table 1**), key vegetation condition criteria are obviously altered, hence the potential to reinstate regenerative landscape processes and over time the original native vegetation is likely to have been variously diminished.

A powerful ecological analysis system 'VAST', which stands for 'vegetation, assets, states and transitions' has been developed to account for the interactions between land management regimes and their effects on transforming ecological function (Fn), vegetation structure (St), species composition (Co). As the name implies, VAST accounts for changes in the condition of vegetation, that is the asset, by assigning a condition state or class to that asset and by tracking transitions between states over time. Tracking vegetation transitions gives different information from mapped vegetation states: it provides decision-makers with information on changes and trends in the resource base due to the environmental and anthropogenic changes, allows land managers to monitor the outcomes of management interventions, and indicates to all stakeholders the link between use and management and observed changes in condition over time.

Using this approach, change is assessed at the site-level (i.e. local soil–landscape unit) over time relative to a fully natural reference [7, 13, 16]. It is also relevant to monitoring and reporting on the transformation at the landscape or regional scale using measures of ecological function (Fn), vegetation structure (St), species composition (Co) over time [7, 16].

To understand the effects of land management practices on key criteria and indicators of vegetation condition over time, it is necessary to compile a systematic chronology of observations of the effects of land management practices. Fundamental to this approach is that site-level qualitative and quantitative data and information are compiled from relevant sources. Where these data are assessed as fit-for-the-purpose (i.e. comprehensive, relevant, and adequate), for the assessment of spatial and temporal changes and trends in condition it forms part of the site history. Where on-ground observations of ecological attributes are determined not fit-for-the-purpose, fine-scale multi-temporal remote sensing and environmental modelling may be used to develop the requisite attribute data to populate criteria and indicators of vegetation condition.

A metric-based system is used to classify and map vegetation 'condition states' which represent a gradient of modification states that are caused by the effects that land use and/or land management practices on compositional, structural and functional characteristics of the site. The aggregate extent to which indicators remain intact or have been modified or removed altogether determines the score out of an overall index 100%. The total index of 100% is divided equally into six classes (**Table 2**). This framework, combined with criteria and indicators, is relevant for assessing the degree of landscape transformation and for assessing the potential to reinstate regenerative landscape processes.

Increasing vegetation modification	Vegetation condition classes	Characteristics of the vegetation
	0 Naturally bare	Areas where native vegetation does not naturally persist, and recently naturally disturbed areas where native vegetation has naturally been entirely removed (i.e. subject to primary succession)
	I Residual	Native vegetation community structure and composition, with regenerative capacity intact—no significant perturbation from land use/land management practices
	II Modified	Native vegetation community structure, composition and regenerative capacity more or less intact, perturbed by land use/land management practices such as intermittent low intensity grazing
	III Transformed	Native vegetation partly removed but community structure, composition has been significantly altered by land use/land management practices. Regenerative capacity usually minimally modified or moderately modified
	IV Largely replaced and plant community is adventive	Native vegetation largely replaced by invasive native and/or exotic plant species (commonly areas abandoned or burnt). Regenerative capacity usually moderately to heavily modified
	V Replaced and managed for intensive production	Native vegetation completely removed and replaced with intensive agriculture: rain-fed broad acre crops, feed lots, horticulture, irrigation agriculture and long or short rotation timber production. Regenerative capacity usually moderately to heavily modified
	VI Replaced with man-made structures/ infrastructure	Native vegetation completely removed and replaced with settlements and cultural features—e.g. buildings, roads, water reservoirs; gardens, parks and amenity plantings. Regenerative capacity usually moderately to heavily modified

Based on: [7, 16].

Table 2. Vegetation condition classes and commonly observed characteristics.

The method for assessing changes in condition over time involves compiling data and information on the effects of land management practices on 10 criteria and 22 indicators of vegetation structure, composition and function (**Table 3**). As a condition framework that needs to be populated, these indicators also provide a rationale for measurement, monitoring and reporting change in condition at sites and landscapes [13, 17]. A brief overview of the method is documented here.

Level 1	Level 2	Level 3
Key indicators	**Key functional, structural and composition criteria**	**Condition components**
1. Area/size of fire	Fire regime	Regenerative capacity
2. Interval between fires		
3. Plant available water holding capacity	Soil hydrology	
4. Groundwater dynamics		
5. Effective rooting depth of the soil profile	Soil structure	
6. Bulk density of the soil through changes to soil structure or soil removal		
7. Nutrient stress—rundown (deficiency) relative to reference soil fertility	Soil nutrient status	
8. Nutrient stress—excess (toxicity) relative to reference soil fertility		
9. Organisms responsible for maintaining soil porosity and nutrient recycling	Soil biological status	
10. Surface organic matter, soil crusts		
11. Reproductive potential of overstorey structuring species	Reproductive potential	
12. Reproductive potential of understorey structuring species		
13. Overstorey top height (mean) of the plant community	Overstorey structure	Vegetation structure
14. Overstorey foliage projective cover (mean) of the plant community		
15. Overstorey structural diversity (i.e. a diversity of age classes) of the stand		
16. Understorey top height (mean) of the plant community	Understorey structure	
17. Understorey ground cover (mean) of the plant community		
18. Understorey structural diversity (i.e. a diversity of age classes) of the plant		
19. Densities of overstorey species functional groups	Overstorey composition	Species composition
20. Richness—the number of indigenous overstorey species relative to the number of exotic species		
21. Densities of understorey species functional groups	Understorey composition	
22. Richness—the number of indigenous understorey species relative to the number of exotic species		

Based on: [19].

Table 3. List of indicators, criteria and components of vegetation condition. Change is assessed relative to an assumed pre-European[1] benchmark. A fourth level results in a vegetation status or transformation index derived by adding the weighted scores from level 3.

. [1] That is, prior to the European settlement of Australia

The 22 indicators (**Table 3**) expand on the commonly used indicators of native vegetation condition: (i) species composition, (ii) vegetation structure and (iii) function [6, 7, 18]. The method comprises an index hierarchy: how to definite a site, what data and information are essential, and the system for scoring and weighting the effects of land management practices using indicators. Multi-criteria analysis (MCA) is used to compile, analyse and report the results including the objective and decision criteria, assemble data inputs, explore and combine data, develop site-based assessments, and review and report [13].

Four levels of assessment underpin the analysis of vegetation condition to enable reporting over time (**Table 3**):

Level 1. Twenty-two *indicators* represent the key vegetation and ecological characteristics that are affected by land management practices;

Level 2. Ten *criteria*. For example, regenerative capacity is affected by soil nutrient state, fire regime and soil biological state and reproductive potential of the overstorey for each year;

Level 3: Three *condition components*: species composition, vegetation structure and regenerative capacity (ecological function); and

Level 4: A *vegetation transformation index* assembled from the scores of the three components.

This hierarchy allows the user to assess change at all levels over time relative to a reference state. Assessments are based on recorded observations and attribute measurements of the criteria and indicators (their magnitude, frequency, duration, seasonality) in parallel with land management practices.

It is assumed that each soil–landscape unit had a homogeneous plant community prior to modern human interventions. Scores for species composition, community structure and regenerative capacity for each site are calculated for different time periods. The scores are weighted approximately 20:30:50 to reflect their relative importance in maintaining resilience and integrity of plant communities, and summed to give a total vegetation transformation index (expressed as a percentage). The index is put into one of five score classes: unmodified (80–100%), modified (60–80%), transformed (40–60%), replaced/adventive (20–40%) and replaced/managed (0–20%). The timeline of changes in the plant community at a site is then set alongside historical and contemporary records, and relationships are established via a set of ecological attributes.

Many land management practices are used to maintain, enhance, restore, degrade, remove, replace or convert vegetation, and to routinely monitor the effects of these practices on plant communities. Relevant data and information on the breadth of land management regimes and land management practices and their effects on vegetation condition at sites over time must be systematically collected and compiled if change is to be monitored. The approach described above provides a format for systematic assemblage and correlation of historical and environmental records at sites. The resulting insight into the origins of the current status of sites can then be used to inform decision-making about restoration and regeneration.

http://dx.doi.org/10.5772/62874

Two case studies show the compilation of data and information on the effects of land management on 22 key ecological characteristics of ecosystem structure, composition and function.

3. Case studies

3.1. Case study 1: condition inferred from land use mapping for Australia

Land use attributes in the 2001/2002 Land Use of Australia, Version 3 were reassigned condition classes for the 2001/2002 reporting year based on knowledge of the interactions between land management regimes and the condition criteria affected by each land use class [20]. It is worth noting that this land use dataset has a time series for the years 1992/1993, 1993/1994, 1996/1997, 1998/1999, 2000/2001 and 2001/2002. While the method for reassigning vegetation condition classes to land use classes for each grid cell could be applied to each of the years in the time series, only 2001/2002 is presented here for illustrative purposes.

The 2001/2002 Land Use of Australia is a compiled land use dataset with geographical coordinates referred to GDA94. Land use types were compiled into grids using ARC/INFO (Trademark). **Figure 1** includes variable grid cell sizes; Australia 0.01° cell size and Tasmania were replaced by catchment scale land use mapping 1:25,000 and 1:100,000 to more accurately represent forests managed for timber production.

Primary (upper case) and secondary (lower case) land use types
1. AGRICULTURE Crops
2. AGRICULTURE Horticulture
3. AGRICULTURE Irrigation
4. FORESTS Agroforestry
5. FORESTS Natural
6. FORESTS Plantation
7. HUMAN ENVIRONMENT
8. LAND Conservation
9. LAND Conservation Reserve
10. WATER Lakes

Primary (upper case) and secondary (lower case) land use types
11. WATER Wetlands

Table 4. Primary (upper case) and secondary (lower case) land use types defined in the 2001/2002 Land Use of Australia, Version 3 [20] which were cross-tabulated with the six land use classes described in the Australian Land Use and Management classification [14].

The extent of non-agricultural land uses are based on existing digital maps covering four themes: protected areas, topographic features, tenure and forest. The extent of agricultural land uses was modelled using the Australian Bureau of Statistics' agricultural censuses and surveys and spatially extended with the aid of Advanced Very High Resolution Radiometer (AVHRR) satellite imagery with ground control data. A detailed data lineage of the 2000/2002 dataset is available http://www.agriculture.gov.au/abares/aclump/Documents/Nat_Luse_Metadata.pdf

The 2001/2002 land use dataset defined a hierarchy of land use attributes including primary classes (upper case) and secondary classes (lower case) as well as tertiary labels (not shown), **Table 4**. These land use attributes and the five land use classes described above are complementary.

Land use attributes (primary, secondary and tertiary attributes) in the 2001/2002 Land Use of Australia were reassigned into seven condition classes based on the knowledge of the interactions between land management regimes and the condition criteria generally affected by each land use class [20]. For example, the *LAND Conservation Reserve* (i.e. land use class 1 Conservation and natural environments in **Table 1**) was assigned a VAST Residual (I) class (**Table 2**). The resulting condition classes are mapped in **Figure 1**, and the classes are described in **Table 2**.

It is acknowledged that relying on the reassignment of land use attribute datasets to derive a vegetation condition dataset provides an incomplete representation of changes in native vegetation condition [20]. To more fully monitor and report, these changes that they used several input datasets to infer the effects of land management on vegetation condition. The input datasets were:

- Native vegetation extent 2004
- Biophysical Naturalness 1995
- Wilderness of potential national significance 1995, 2000
- Australian Collaborative Land Use Mapping Program land use 2000/2001
- MODIS Land Cover Type 2004 (MOD12Q1), International Geosphere–Biosphere Programme Classification.

Condition was modelled by using the above spatial/temporal datasets (derived from 1995 to 2006) and an expert model that compared the effects of land use and land management practices on vegetation with an assumed pre-1750 vegetation condition benchmark.

http://dx.doi.org/10.5772/62874

Improvements in the mapping of vegetation condition could be achieved by including changes in the extent of native vegetation, changes in the extent, severity and seasonality of unplanned fire (i.e. bushfires) and changes in the cover and density of invasive naturalised pasture species [20].

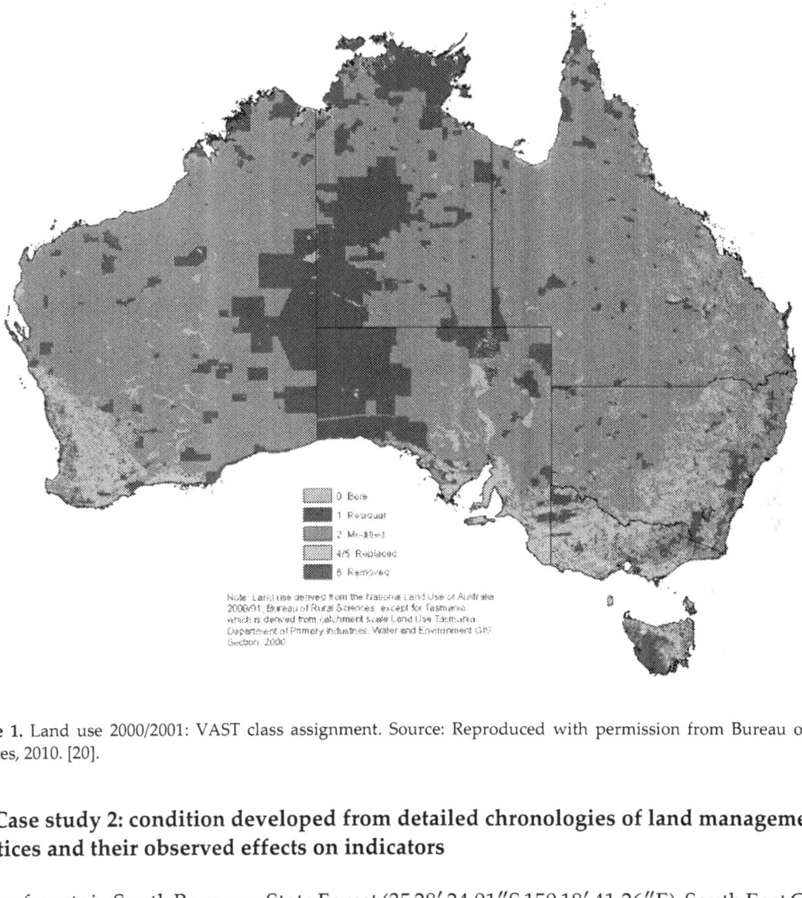

Figure 1. Land use 2000/2001: VAST class assignment. Source: Reproduced with permission from Bureau of Rural Sciences, 2010. [20].

3.2. Case study 2: condition developed from detailed chronologies of land management practices and their observed effects on indicators

Native forests in South Brooman State Forest (35 28′ 24.01″S 150 18′ 41.26″E), South East Corner bioregion, Australia, have a long history of management for wood production. The tall open Eucalypt open forest (*Corymbia maculata*) in this area has been managed using two levels of management intervention (**Figure 1**) [21]. In 1880, the site was minimally modified by selectively harvesting the larger trees and left to recover. Between 1944 and 1959, the site was intensively managed for timber production refer to 'C' in **Figure 2** where the site was picked over for high-quality sawlogs, and since 2004, the site has been managed to encourage regeneration. Throughout the period from first intervention in the 1870s to 2011, changes in species composition and regenerative capacity have been minimal. These interventions have focussed on the vegetation structure, by harvesting trees of different age classes and since 1960 by improving the structural diversity of the forest.

The changes shown in **Figure 2** for spotted gum (*Corymbia maculata*) forest in South Brooman State Forest, New South Wales.

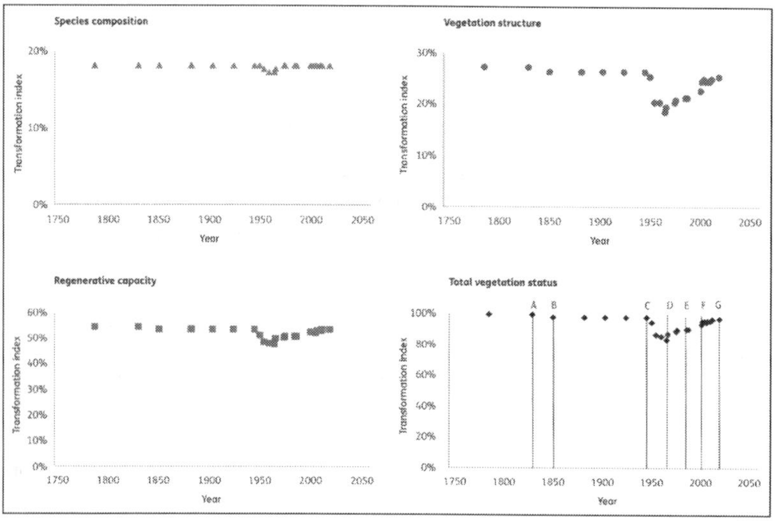

Figure 2. Changes in spotted gum forest, South Brooman State Forest, Batemans Bay, New South Wales, Australia. Key management practices: A = indigenous management; B = site picked over for high-quality sawlogs, fire suppressed and/or excluded; C = site again picked over for high-quality sawlogs; D = sawlogs harvested over 85% of site, removing 50% of canopy; E = site rehabilitated naturally; F = wildfire burnt 100% of site; G = site rehabilitated naturally. Source: http://aceas.org.au/portal Vegetation Transformation Study Sites, South Brooman State Forest Site Vegetation Transformation Details [21].

4. Discussion

The above approaches to monitoring and reporting vegetation condition differ from other frameworks for monitoring and reporting vegetation condition [5], which are mostly biodiversity-based frameworks that have been widely applied [22]. Those biodiversity-based frameworks consider vegetation attributes in isolation from, or independently of, land management practices and only consider *post hoc* the effects of changes in land management, in order to explain observed patterns of environmental information.

Another landscape condition assessment framework is the landscape functional analysis (LFA), which focuses on attributes including landform, soil, hydrology and nutrient status. LFA was developed to help land managers restore ecological function in degraded ecosystems [23]. The LFA framework considers attributes of native vegetation (i.e. structure and composition) as a means to an end, being indicators of healthy functioning landscapes. Hence, detailed species lists are not necessary for LFA.

Biodiversity- and landscape condition-based assessment frameworks [22, 23] tend to collate land use and land management to interpret the current condition state but generally do not thoroughly and systematically analyse the effects of historic and contemporary land management practices on criteria and indicators of vegetation condition. They therefore provide little operationally relevant information that is at a scale that is useful to land managers.

The approaches described in this chapter do not seek to supersede nor replace these existing frameworks for assessing condition. Rather, they aim to provide a more integrated system for evaluating anthropogenic effects on the condition of native vegetation communities over time.

The development and application of criteria and indicators of vegetation condition can show which sites and landscapes have not been modified or only minimally modified. This information can help land managers who make decisions about land management practices and to identify areas which have been rehabilitated towards a threshold state as a result of a change/s in land management regimes. Other applications of condition-based decision-making include setting goals or targets for improving vegetation condition across sites and landscapes; identifying and selecting priority areas for restoration and regeneration; or monitoring and reporting the outcome of changed land use and management. It is critical to understand which criteria and indicators of the extent and condition of the native vegetation have been modified or changed relative to a reference state and to what extent, and over what time period [24, 25]. Monitoring and reporting the transformation pathways of native vegetation at site and landscape scales including processes of replacement, removal and recovery are vital but is often poorly implemented and practiced by land managers and scientists alike. Where areas are needed for biodiversity conservation, selecting those areas for treatment that are least modified is likely to result in a more rapid response, following a change in land management practices relative to the reference state.

Modifications and changes due to the anthropogenic interventions occur at the site and landscape scales and change may be slow and almost imperceptible or rapid and both may involve dramatic modification of key vegetation and/or ecological attributes.

Given this setting, the challenge for decision-makers is to develop practical, meaningful and accessible tools and systems for tracking these changes, to monitor and report anthropogenic site and landscape transformations and to link these observed ecological manipulations to ecosystem service benefits and dis-benefits effecting human well-being, biodiversity and ecosystem function.

The development of detailed and systematic chronologies of land management practices and their effects on indicators of vegetation condition places equal importance in collating qualitative and quantitative historical records. Because much of these data and information have not been compiled at the site-level, it is fundamental to work with on-ground land managers to understand the current condition state and what transitions have occurred in key criteria and indicators which have led up to that condition state. Changes include loss of soil structure and nutrients, regrowth of vegetation structure and incursions of new species and modified fire regimes. Often at the soil–landscape level, it is the land manager (and others) who have observed changes in criteria and indicators over time as a result of their own changes in land management practices.

Where detailed on-ground observations of changes in the above indicators are absent in the chronological record, remote sensing and environmental modelling may be used to develop the requisite essential environmental measures and derived attribute data to populate criteria and indicators of vegetation condition.

Effecting future changes in condition state must also include the land manager. Access to synthesised information generated by compiling a detailed chronology of land management practices and their effects helps land managers to better understand what criteria and indicators they are able to modify and change by changing their land management regimes using appropriate and fit-for-purpose land management practices. Land managers and other decision-makers can use the above approach to gain valuable insights for adaptive management. Decision-makers can in turn use that information for a number of purposes, for example:

1. to enable more reliable identification and selection/s of soil–landscape areas where the structural and compositional have been minimally or moderately modified and where the ecological functions have not been damaged,

2. to inform discussions with land managers on what changes in land management practices and land management regimes that are likely to yield long-lasting changes in condition relative to a reference estate; and

3. to guide the collection of monitoring data and information relevant to the land manager and other interested parties including land management practices and attributes of condition.

5. Conclusions

The case studies above illustrate transformation pathways including process of replacement, removal and recovery of native vegetation and provide information about how sites and landscapes have been transformed, for example what, where and when an intervention/s was/were used and what was the observed result. The systematic analytical framework used in these case studies to assess changes in vegetation condition offers certainty for land managers and planners through providing consistent results at sites and landscapes over time.

The above discussion has shown that monitoring and reporting vegetation condition information can be derived by direct on-ground measurement and population of criteria and indicators of vegetation condition or by inferring condition information from available land use maps and a knowledge of land management regimes and criteria that are deliberately or inadvertently changed or modified using land management practices. It is worth noting that the same 10 criteria (**Tables 1** and **2**) are used in both approaches. Condition information

http://dx.doi.org/10.5772/62874

generated from land use maps alone will, however, only enable generalised understandings, whereas developing a systematic chronology of land management practices linked to on-ground direct measurement and assessing criteria and indicators of vegetation condition will provide decision-makers with greater reliability and flexibility and be relevant to multiple applications.

Where land planners and decision-makers require information on the status, change and trend of these impacts on vegetation types for environmental reporting, land use trade-offs and to help assess future land use scenarios, direct on-ground measurement and population of criteria and indicators of vegetation condition should be preferred.

The two approaches described above can be used to inform land managers and other decision-makers to establish and link patterns in the use and management of native vegetation and the transformation of vegetated landscapes. Only the second approach, of developing detailed systematic chronologies of land management and their effects on indicators of vegetation condition, can help answer the following key questions: What is the condition of the native vegetation at a site and landscape over time based on the impacts on land management on indicators of structure, composition and function? How can I monitor and report the condition of native vegetation resulting from management interventions? As a land manager, how can I use knowledge of the impacts of land management on indicators of structure, composition and function to improve the condition of the native vegetation of my site and landscape?

Acknowledgements

Jake Gillen, Mark Parsons and Shane Cridland provided comments on an early draft.

Author details

Richard Thackway

Address all correspondence to: rthackway@netspeed.com.au

School of Geography, Planning and Environmental Management, The University of Queensland, Brisbane, Australia

References

[1] Yapp G, Walker J, Thackway R. Linking vegetation type and condition to ecosystem goods and services. Ecological Complexity 2010; 7(3):292–301. doi:10.1016/j.ecocom.2010.04.008

[2] Trudgill, ST. Soil and vegetation systems; contemporary problems in geography. Oxford, Clarendon Press. p211.ISBN 0198741391. 1988.

[3] Woollacott A, Adcock M, Allen M, Evans R, Mackinnon, A. Overview: the Making of the Modern World (1750–1918). Cambridge: Cambridge University Press; 2010.

[4] Blainey G. The Story of Australia's People. The rise and fall of ancient Australia: volume 1. Melbourne: Viking. 2015

[5] EMR (Ecological Management and Restoration). Special Issue: native vegetation condition mapping 2006; 7 Suppl 1:S1–S80. http://www.blackwell-synergy.com/loi/emr

[6] Gibbons P, Freudenberger D. An overview of methods used to assess vegetation condition at the scale of the site. Ecological Management and Restoration 2006;7:S10–S17. doi:10.1111/j.1442-8903.2006.00286.x

[7] Thackway R, Lesslie R. Describing and mapping human-induced vegetation change in the Australian landscape. Environmental Management 2008;42:572–590. doi:10.1007/s00267-008-9131-5

[8] Lawley V, Lewis M, Clarke K, Ostendorf B. Site-based and remote sensing methods for monitoring indicators of vegetation condition: an Australian review. Ecological Indicators 2016;60:1273–1283. doi:10.1016/j.ecolind.2015.03.021

[9] State of the Environment Committee. Native vegetation, Australia State of the Environment 2011. Independent Report to the Australian Government Minister for Sustainability, Environment, Water, Population and Communities. Canberra; 2011 [Internet] [cited 17 January 2016]. Available from http://www.environment.gov.au/science/soe/2011-report/5-land/2-state-and-trends/2-3-vegetation#ss2-3-1

[10] Montreal Process Implementation Group for Australia and National Forest Inventory Steering Committee. Australia's State of the Forests Report. Canberra: ABARES; 2013 [Internet] [cited 17 January 2016]. Available from http://www.daff.gov.au/ABARES/forestsaustralia/Documents/2013-web2.pdf

[11] Hnatiuk R, Thackway R, Walker J. Vegetation. In: National Committee for Soil and Terrain Information (Ed). Australian Soil and Land Survey: Field Handbook (3rd edition). Melbourne: CSIRO Publishing; 2009.

[12] Lesslie R, Mewett J. Land Use and Management: the Australian Context. Australian Bureau of Agricultural and Resource Economics and Sciences, Research Report 13.1, Canberra; 2013.

[13] Thackway R, Specht A. Synthesising the effects of land use on natural and managed landscapes. Science of the Total Environment 2015;526:136–152. doi:10.1016/j.scitotenv.2015.04.070

[14] Australian Bureau of Agricultural and Resource Economics and Sciences. Guidelines for land use mapping in Australia: principles, procedures and definitions. A Techni-

cal Handbook Supporting the Australian Collaborative Land Use and Management Program 4th edition, 2011. Canberra: Australian Bureau of Agricultural and Resource Economics and Sciences; 2011.

[15] Bradstock RA, Gill AM, Williams JE (Eds). Flammable Australia: The Fire Regimes and Biodiversity of a Continent. Cambridge: Cambridge University Press; 2002.

[16] Thackway R, Lesslie R. Reporting vegetation condition using the Vegetation Assets, States and Transitions (VAST) framework. Ecological Management and Restoration 2006;7(S1):S53–S62. doi:10.1111/j.1442-8903.2006.00292.x

[17] Thackway R. Applying a system for tracking the changes in vegetation condition to Australia's forests. In: Brown AG, Wells KF, Parsons M, Kerruish CM (Eds). Managing our Forests into the 21st Century. Proceedings of National Conference, Institute of Foresters of Australia, Canberra, ACT, Australia, 4–7 April 2013, 2013; pp. 79–91. Available from http://forestryconference.org.au/program/program-abstracts/ifa-2013-conf-papers-thackway

[18] Noss RF. Indicators for monitoring biodiversity: a hierarchical approach. Conservation Biology 1990;4(4):355–364. doi:10.1111/j.1523-1739.1990.tb00309.x

[19] Yapp GA, Thackway R. Responding to Change—Criteria and Indicators for Managing the Transformation of Vegetated Landscapes to Maintain or Restore Ecosystem Diversity, Biodiversity in Ecosystems—Linking Structure and Function, Dr Juan A. Blanco (Ed), InTech, 2015; ISBN: 978-953-51-2028-5. Available from: http://www.intechopen.com/books/biodiversity-in-ecosystems-linking-structure-and-function/responding-to-change-criteria-and-indicators-for-managing-the-transformation-of-vegetated-landscapes

[20] Lesslie R, Thackway R, Smith, J. A National-Level Vegetation Assets, States and Transitions (VAST) Dataset for Australia (version 2). Canberra: Bureau of Rural Sciences; 2010.

[21] Thackway R. Transformation of Australia's vegetated Landscapes, South East bioregion, South Brooman State Forest, NSW. ACEAS, University of Queensland; 2012. doi:10.4227/05/5088F50BD771D. Available from http://dx.doi.org/10.4227/05/5088F50BD771D

[22] Parkes D, Newell G, Cheal D. Assessing the quality of native vegetation: the 'habitat hectares' approach. Ecological Management and Restoration 2003;4:S29–S38. doi: 10.1046/j.1442-8903.4.s.4.x

[23] Tongway D, Ludwig J. Restoring Disturbed Landscapes Landscape Ecology. USA: Island Press; 2011.

[24] SERISPWG (Society for Ecological Restoration International Science and Policy Working Group). The SER International Primer on Ecological Restoration. Tucson: Society for Ecological Restoration International; 2004 [Internet] [cited 17 January 2016].

Available from http://www.ser.org/resources/resources-detail-view/ser-international-primer-on-ecological-restoration#8

[25] Hobbs RJ, Walker LR, Walker J. Integrating restoration and succession, In: Walker LR, Walker J, Hobbs RJ (Eds). Linking restoration and ecological succession. New York: Springer; 2007; pp. 168–179.